GLOBAL ECOLOGY

The magnitude and rapidity of global environmental degradation threatens the perpetuation of life on Earth. Yet our understanding of biospheric change may not be sophisticated enough to adopt the long-term management strategies necessary to put modern civilization on a sustainable basis.

Global Ecology evaluates the current state of knowledge concerning biospheric change, recognizing the limits of scientific studies and quantitative modelling, and analysing the weaknesses and uncertainties of our environmental understanding. A critical assessment of existential needs discusses the levels of food, energy, water and materials necessary to support a decent quality of life.

Global Ecology juxtaposes the encouraging potential for effective solutions with the numerous environmental, technical and social obstacles that limit and counteract efforts to improve our management of natural resources and reduce environmental degradation. With a strong plea to preserve flexibility of adaptive actions in managing the transition to a more sustainable society, the author leads the reader to a greater understanding of our ability to manage the effects of biospheric change.

Vaclav Smil is Professor of Geography at the University of Manitoba, Canada.

GLOBAL ECOLOGY

Environmental change and social flexibility

Vaclav Smil

London and New York

First published 1993
by Routledge
11 New Fetter Lane, London EC4P 4EE

Simultaneously published in the USA and Canada
by Routledge
29 West 35th Street, New York, NY 10001

© 1993 Vaclav Smil

Typeset in Sabon by Solidus (Bristol) Ltd, Bristol, England
Printed and bound in Great Britain by
Mackays of Chatham PLC, Chatham, Kent

British Library Cataloguing in Publication Data
A catalogue record for this book is available from the British Library

Library of Congress Cataloging in Publication Data
Smil, Vaclav.
Global ecology : environmental change and social flexibility /
Vaclav Smil.
p.   cm.
Includes bibliographical references (p.   ) and index.
ISBN 0–415–09885–8. — ISBN 0–415–09886–6 (pbk.)
1. Environmental policy.   2. Sustainable development.   I. Title.
GE170.S65   1994
363.7—dc20   93–9856
                                                                                  CIP

ISBN 0–415–09885–8 (hbk)   0–415–09886–6 (pbk)

Whilst the multitude of men degrade each other, and give currency to desponding doctrines, the scholar must be a bringer of hope and must reinforce man against himself.

Ralph Waldo Emerson *The Method of Nature* (1841)

CONTENTS

FIGURES

PREFACE

'Society is a wave. The wave moves onward, but the water of which it is composed does not,' wrote Emerson in his essay on self-reliance (1860). And so we have had, starting in the early 1970s, another busy period of questioning the human prospect. Concatenation of the end of a generation of rapid economic growth, growing anxieties about resource scarcity and environmental pollution, rising energy costs, record low grain stocks and several famines made the early 1970s into a time of great doubt. 'There is a question in the air', noted Robert Heilbroner in 1974, 'more sensed than seen . . . a question that I would hesitate to ask aloud did I not believe it existed unvoiced in the minds of many: Is there hope for man?' His answer, although qualified, was 'No'. At the same time the Club of Rome predicted 'a rather sudden and uncontrollable decline in both population and industrial capacity' during the next century, a conclusion based on computer models of *The Limits to Growth* (Meadows *et al.* 1972).

The second half of the 1970s and the early 1980s did not bring much cheer: energy prices rose even higher, global economy went into a recession, superpower relations deteriorated. In the panic gold prices rose sharply and *The Bulletin of the Atomic Scientists* pushed its doomsday clock just three minutes to midnight. The 1980 *Global 2000 Report to the President* was hardly more comforting than *The Limits* published in 1972. Not surprisingly, there arose a small wave of optimistic antidotes portraying a future of unlimited prospects: *The Resourceful Earth* (Simon and Kahn 1984) was its most eloquent example. Yet its critical reading shows this riposte to carry no fewer biases and errors than the earlier prophecies of doom. Still, the world of the mid-1980s seemed a more hopeful place with its easing of superpower tensions, rising stock markets, lower energy prices and encouraging regional moves toward food self-sufficiency, democracy and end of armed conflicts.

These were short-lived reversals. The end of the decade and the beginning of a new one became, once again, gloomy, and not even the demise of the Soviet empire could dispel these feelings. Growing burdens of overspending in rich nations and of the huge debt in the poor ones, hopeful but also potentially dangerous disintegration of one superpower and a discomfiting malaise of the other, advancing pandemics of AIDS, and the resurgence of atavistic ethnic conflicts now unfold against a background of unprecedented concerns about the magnitude and rapidity of global environmental change.

Its ingredients have become well known. Massive burning of Amazonian forests, indiscriminate logging in South-east Asia and food and fuel needs of Africa's fast-growing population are destroying tropical rain forests, the Earth's richest repositories of biodiversity. Soil erosion, desertification, improper irrigation, inadequate recycling of organic matter and excessive use of farm chemicals are reducing the extent and the quality of arable lands. The Antarctic ozone hole and a possibility of its Arctic duplicate are causing fears of extensive damage to crops, animals and human health. And the anticipation of rapid climatic change is moving nations toward a formulation of global co-operative policies designed to forestall the burdens of reduced harvests, declining economies and masses of environmental refugees.

In this book I will first appraise our understanding of these, and other, threats to the perpetuation of life on the Earth (Chapter 1). Then I will look at fundamental existential requirements (Chapter 2) and proceed with a detailed assessment of all major environmental concerns (Chapter 3). The next two chapters (Chapters 4 and 5) will be devoted to a critical review of numerous options we have to deal with this worrisome deterioration – and no less numerous complications which will hinder such adaptations. This sequence is deliberate: a sensible appraisal of our capabilities must resonate with tensions and contradictions.

After finishing Chapter 4 reviewing the possibilities of effective advances and solutions, a reader should feel justifiably confident that the combination of human inventiveness and determination offers a way out from our predicament. But after concluding Chapter 5 which details major obstacles on this path a reader may have serious doubts if this promise can be realized. This disconcerting contrast is undoubtedly exasperating and depressing – but to believe that we can bridge the gap between the two realities is not a matter of foolish faith but rather a cause for affirmation of hope and for a determined pursuit of creative thinking and unprecedented solutions (Chapter 6). Appreciation of the gap is essential in order to act rationally – to realize the enormity of the task ahead and to marshal requisite responses.

In portraying this quintessentially *yin-yang* quality of the challenge I will use no elaborate analytical techniques, no complicated forecasting models, just simple observations, commentaries, probes and questions returning to some great constants of our successes and failures. Selection of examples, the choice of arguments and the very way of overall presentation also reflect personal inclinations and preferences. I am well aware that such an approach cannot be devoid of biases, lapses, and contradictions. But a clear personal voice does not make the book normative: I am not prescribing any definite solutions, I am merely showing the dimensions of essential requirements, pointing approvingly in some directions and arguing for preservation of maximum flexibility.

1

ELUSIVE UNDERSTANDING

WHY THERE IS SO MUCH UNCERTAINTY
ABOUT ENVIRONMENTAL CHANGE

Perhaps too the difficulty is of two kinds, and its cause is not so much
in the things themselves as in us.

Aristotle *Metaphysics*

Effective action may not require perfect understanding, but a sensible, realistic grasp is essential. After generations of intensifying scientific research and management experience we have such a knowledge to ease our way through many difficult environmental challenges – but we still lack it in a much larger number of cases, including most of the matters of critical global importance. Moreover, we will not be able to resolve most of the uncertainties during the next one or two generations, that is, during the time of increasing pressures to adopt unprecedented preventive and remedial actions.

Undoubtedly the greatest obstacle to a satisfactory understanding of the Earth's changing environment is its inherent complexity, above all, the intricate and perennially challenging relationship of parts and wholes, a great *yin-yang* of their dynamic interdependence. But this objective difficulty is greatly complicated by subjective choices and biases of people studying the environment: overwhelmingly, they favour dissection and compartment-alization to synthesis and unification, they rely excessively on modelling which presents too frequently a warped image of reality, and they are not immune to offering conclusions and recommendations guided by personal motives and preferences.

This makes it very difficult to separate clear facts from subtle biases and wilder speculations even for a scientist wading through a flood of publications on fashionable research topics. Policy-makers often select inter-pretations which suit best their own interests, and the general public is exposed almost solely to findings with a sensationalizing, catastrophic bias. To come up with generally acceptable sets of priorities for effective remedies is thus very difficult – and bureaucratization and politicization of environ-mental research and management make it even harder to translate our imperfect understanding into sensible actions. Before addressing these complications, I will first review the genesis of our environmental understanding and the emergence of environmental pollution and degradation as matters of important public concern.

MODERN CIVILIZATION AND THE ENVIRONMENT

Yet we can use nature as a convenient standard, and the meter of our rise and fall.

Ralph Waldo Emerson, *The Method of Nature* (1841)

Shortly after the beginning of the twentieth century the age of the explorer was coming to the close. Yet another generation was to elapse before the last isolated tribes of New Guinea were contacted by the outsiders (Diamond 1988), but successful journeys to the poles – Robert Peary's Arctic trek in 1909 and Roald Amundsen's Antarctic expedition in 1911 – removed the mark of *terra incognita* from the two most inaccessible places of the Earth (Marshall 1913).

Four centuries of European expansion had finally encompassed the whole planet. But this end of a centuries-old quest was not an occasion for a reflection on limits and vulnerabilities of the Earth: such feelings were still far in the future, in the time of Apollo spacecrafts providing the first views of the blue-white Earth against the blackness of the cosmic void.

For an overwhelming majority of people life at the beginning of the early twentieth century differed little from the daily experience of their ancestors two, ten, or even twenty generations ago. They were subsistence peasants, relying on solar energies and on animate power to cultivate low-yielding crops. Even the industrializing countries of Europe and North America were still largely rural, horses were the principal field prime movers, and there were hardly any synthetic fertilizers. Environmental degradation and pollution accompanied the advancing urbanization and mass manufacturing based on coal combustion, but these effects were highly localized.

Grimness of polluted industrial cityscapes was an inescapable sign of a new era – but not yet a matter of notable public, governmental or scientific concern. Study of the environment had little to do with changes brought by man. The earliest attempt at a grand synthesis of man's impact on natural environments (Marsh 1864) was so much ahead of its time that it had to wait almost another century for its successor (Thomas 1956). But foundations were being laid for a systematic understanding surpassing the traditional descriptions. During the nineteenth century most of the available information about the Earth's environment was admirably systematized in such great feats of advancing science as writings of Alexander von Humboldt (1849), Henry Walter Bates (1863) and Alfred Russel Wallace (1891), or in the maps of the British Admiralty. But our understanding of how the environment actually works was only beginning.

Rapidly advancing biochemistry and geophysics were laying the foundations for understanding the complexities of grand biospheric cycles. By the late 1880s Hellriegel and Wilfarth (1888) identified the nitrogen-fixing symbiosis between legumes and *Rhizobium* bacteria, by the late 1890s Arrhenius (1896) published his explanation of the possible effects of atmospheric CO_2 on the ground temperature and Winogradsky isolated nitrifying bacteria (Winogradsky 1949). At the same time, Charles Darwin's (1859) grand syntheses turned attention to the interplays between organisms and their surroundings, and physiologists offered revolutionary insights into the nutritional needs of plants, animals and men (Atwater 1895). Still, the blanks dominated, and the absence of highly sensitive, reproducible analytical methods precluded any reliable monitoring of critical environmental variables.

Wilhelm Bjerknes set down the basic equations of atmospheric dynamics in 1904 – but climatology remained ploddingly descriptive for a few more decades (Panofsky 1970). Need for an inclusive understanding of living systems was in the air – but ecological energetics was only gathering its first tentative threads (Martinez-Alier 1987). And environmental studies everywhere retained an overwhelmingly local and specific focus and a heavily descriptive tilt with quantitative analyses on timid sidelines.

Everyday treatment of the environment did not show any radical changes – but new attitudes started to make some difference. The closing decades of the nineteenth century saw the establishment of the first large natural reserves and parks, and the adoption of the first environmental control techniques after a century of industrial expansion which treated land, waters and air as valueless public goods. The new century brought the diffusion of primary treatment of urban waste water and the invention of electrostatic control of airborne particulates (Bohm 1982). The two world wars and the intervening generation of economic turmoil were not conducive to greater gains in environmental protection, but a number of ecological concepts found its way into unorthodox economics as well as into some fringe party politics (Bramwell 1989).

Alfred Lotka (1925) published the first extended work putting biology on a quantitative foundation, and Vladimir Ivanovich Vernadsky (1929) ushered the study of the environment on an integrated, global basis with his pioneering book on the biosphere. Arthur Tansley defined the ecosystem, one of the key terms of modern science (Tansley 1935), and Raymond Lindeman, following Evelyn Hutchinson's ideas, published the first quantification of energy flows in an observed ecosystem (Lindeman 1942). While its understanding was advancing, the environment of industrialized countries kept on deteriorating. Three kinds of innovations, commercialized between the world wars, accounted for most of this decline.

Thermal generation of electricity, accompanied by emissions of fly ash, sulphur and nitrogen oxides and by a huge demand for cooling water and the warming of streams, moved from isolated city systems to large-scale integrated regional and national networks (Hughes 1983). Availability of this relatively cheap and convenient form of energy rapidly displaced fuels in most industrial processes and started to change profoundly household energy use with the introduction and diffusion of electric lights, cooking and refrigeration. The last demand was largely responsible for the development of a new class of inert compounds during the 1930s: their releases are now seen as the most acute threat to the integrity of the stratospheric ozone (Manzer 1990).

The automobile industry shifted from workshops to mass production, making cars affordable for millions of people (Lacey 1986) and spreading the emissions of unburned hydrocarbons, nitrogen oxides and carbon monoxide over urban areas and into the countryside which became increasingly buried under asphalt and concrete roads. And the synthesis of plastics grew into a large, highly-energy-intensive industry generating toxic pollutants previously never present in the biosphere, and it introduced into the environment a huge mass of virtually indestructible waste (Katz 1984).

Post-1950 developments amplified these trends and introduced new environmental risks as the world entered the spell of its most impressive economic growth terminated only by the Organization of Petroleum-Exporting Countries' (OPEC's) 1973–4 quintupling of oil prices. In just 25 years global consumption of primary commercial energy nearly tripled, that of electricity went up about eightfold, and there were substantial multiples in demand for all leading metals, as well as in ownership of consumer durables. New environmental burdens were introduced with the rapidly expanding use of nitrogenous fertilizers derived from synthetic ammonia first produced in 1913 (Holdermann 1953; Engelstad 1985), and with the growing applications of the just discovered pesticides (dichloro-diphenyl-trichloro-ethane (DDT) was first used on a large scale in 1944): nitrates in groundwater and streams and often dangerously high pesticide residues in plant and animal tissues.

Environmental pollution, previously a matter of local or small regional impact, started to affect increasingly larger areas around major cities and conurbations and downwind from concentrations of power plants as well as the waters of large lakes, long stretches of streams and coastlines and many estuaries and bays. Better analytical techniques were recording pollutants in the air, waters and biota thousands of kilometres from their sources (Seba and Prospero 1971). Only a few remedies were introduced during the 1950s. Smog over the Los Angeles Basin led to an establishment of the first stringent industrial emission controls in the late 1940s (Haagen-Smit 1970). London's heavy air pollution, culminating in 4000 premature deaths during the city's worst smog episode in December 1952, brought the adoption of the Clean Air Act, the first comprehensive effort to clean a country's air (Brimblecombe 1987).

Electrostatic precipitators became a standard part of large combustion sources, black particulate matter started to disappear from many cities and visibility improved; this trend was further aided by the introduction of cleaner fuels – refined oils and natural gas – in house heating. But as the years of sustained economic growth brought unprecedented affluence to larger shares of North American and European populations, they also spread a growing uneasiness about intensifying air pollution and about degradation of many ecosystems. By the early 1960s these cumulative concerns pushed the environment into the forefront of public attention.

ENVIRONMENT AS A PUBLIC CONCERN

We cannot consider men apart from the rest of the country, nor an inhabited country apart from its inhabitants without abstracting an essential part of the whole.

A.J. Herbertson (1913)

Human impacts became first a matter of widespread public discourse and political import in the United States. Rachel Carson's (1962) influential warning about the destructive consequences of pesticide residues in the environment has been often portrayed as the prime mover of new environmental awareness. In reality, the impulses came from many quarters, ranging from traditional concerns about wilderness preservation to new scientific findings about the fate of anthropogenic chemicals.

Mounting evidence of worsening air pollution was especially important in influencing the public opinion as sulphates and nitrates, secondary pollutants formed from releases of sulphur and nitrogen oxides started to affect large areas far away from the major sources of combustion. Europeans were first to note this phenomenon in degradative lake and soil changes in sensitive receptor areas (Royal Ministry of Foreign Affairs 1971). The decade brought also the first exaggerated claims about environmental impacts. Commoner (1971) feared that inorganic nitrogen carries such health risks that limits on fertilization rates, economically devastating for many farmers, may be needed soon to avert further deterioration. His misinterpretation was soon refuted (Aldrich 1980), but there was no shortage of other targets. Once the interest in environmental degradation began, the western media, so diligent in search of catastrophic happenings, and scientists whose gratification is so often achieved by feeding on fashionable topics, kept the attention alive with an influx of new bad news.

These concerns were adopted by such disparate groups as the leftist student protesters, who discovered yet another reason why they should tear down the *ancien régime*, and large oil companies which advertised their environmental pedigrees in double-page glossy spreads (Smil 1987). Search for the causes of environmental degradation brought theories ranging from an influential identification of the Judaeo-Christian belief (man, made in God's image, set apart from nature and bent on its subjugation) as the principal root of environmental problems (White 1967), to predictable, and even more erroneous, Marxist-Leninist claims putting the responsibility on capitalist exploitation.

White overdramatized the change of the human role introduced by Christianity (Attfield 1983), and ignored both a greater complexity of Christian attitudes and an extensive environmental abuse in ancient non-Christian societies (Smil 1987). Emptiness of ideological claims was exposed long before the fall of the largest Communist empire by detailed accounts of Soviet environmental mismanagement (Komarov 1980).

In the summer of 1970 came the first attempt at a systematic evaluation of global environmental problems sponsored by the Massachusetts Institute of Technology (Study of Critical Environmental Problems 1970). The items were not ranked but the order of their appearance in the summary clearly indicated the relative importance perceived at that time. First came the emissions of carbon dioxide from fossil fuel combustion, then particulate matter in the atmosphere, cirrus cloud from jet aircraft, the effects of supersonic planes on the stratospheric chemistry, thermal pollution of waters and DDT and related pesticides. Mercury and other toxic heavy metals, oil on the ocean and nutrient enrichment of coastal waters closed the list.

Just a month later Richard Nixon sent to the Congress the first report of the President's Council on Environmental Quality, noting that this was the first time in the history that a nation had taken a comprehensive stock of the quality of its surroundings. Soon afterwards his administration fashioned the Environmental Protection Agency by pulling together segments of five departments and agencies: environment entered the big politics. And it entered international politics with the UN-organized Conference on Human Environment in Stockholm in 1972 where Brazilians insisted on their right to cut down all of their tropical forests – and Maoist China claimed to have no environmental problems at all (UN Conference on the Human Environment 1972).

OPEC's actions in 1973–4, global economic downturn and misplaced worries about the coming shortages of energy, turned the attention temporarily away from the environment but new studies, new revelations and new sensational reporting kept the environmental awareness quite high. Notable 1970's concerns included the effects of nitrous oxide from intensifying fertilization on the fate of stratospheric ozone (CAST 1976), carcinogenic potential of nitrates in water and vegetables (NAS 1981), and both the short-term effects of routine low-level releases of radio-nuclides from nuclear power plants and the long-term consequences of high-level radioactive wastes to be stored for millennia (NAS 1980).

And with the economic plight of poor nations worsened by the higher prices of imported oil came the western 'discovery' of the continuing dependence of all rural and some urban Asian, African and Latin American populations on traditional biomass energies (Eckholm 1976) and the realization of how environmentally ruinous such reliance can be in the societies where recent advances in primary medical care pushed the natural increase of population to rates as high as 4 per cent a year. Naturally, attention focused also on other causes of deforestation and ensuing desertification in subtropical countries and soil erosion and flooding in rainy environments: inappropriate ways of farming, heavy commercial logging, often government-sponsored conversion of forests to pastures. Unexpected recurrence of an Indian food shortages and the return of severe Sahelian droughts served as vivid illustrations of these degradative problems.

To this must be added the effects of largely uncontrolled urban and industrial wastes, including releases of toxic substances which would not be tolerated in rich countries, appalling housing and transportation in urban areas, misuse of agricultural chemicals and continuing rapid losses of arable land to house large population increases and to accommodate new industries. Not surprisingly, an ambitious American report surveying the state of the environment devoted much of its attention to the environmental burdens of the poor world (Barney 1980).

In human terms this degradation presented an especially taxing challenge to the world's most populous country. When Deng Xiaoping's 'learning from the facts' replaced Mao Zedong's 'better red than expert', stunning admissions and previously unavailable data made it possible to prepare a comprehensive account of China's environmental mismanagement whose single most shocking fact may be the loss of one-third of farmland within a single generation – in a nation which has to feed a bit more than one-fifth of mankind from 1/15 of the world's arable land (Smil 1984). And the environmental problems of the poor world were even more prominent in yet another global stock-taking in 1982, the Conference on Environmental Research and Priorities organized by the Royal Swedish Academy of Sciences (Johnels 1983).

Environmental mishaps and worries during the 1980s confirmed and intensified several recurring concerns (Ellis 1989). Such accidents as the massive cyanide poisoning in Bhopal, and the news about the daunting effort to clean up thousands of waste sites in the USA, have turned hazardous toxic wastes into a lasting public concern. The Chernobyl disaster in 1986 strengthened the fears of nuclear power. And the discovery of a seasonal ozone hole above Antarctica (Farman *et al*. 1985) revived the worries about the rapid man-induced changes of the atmosphere, a concern further intensified with a new wave of research on the imminent global warming.

Concatenation of these mishaps and worries led to a widespread adoption of environmental change as a major item of national and international policy-making during the late 1980s. Such an eager embrace involves an uncomfortably large amount of posturing, token commitments and obfuscating rhetoric – but it is a necessary precondition for an adoption and eventual enforcement of many essential international treaties. Montreal Protocol of 1987 (and its subsequent perfection), aimed at the reduction and elimination of chlorofluorocarbons, is, so far, the best example of such actions (Wirth and Lashof 1990).

But in a more fundamental way this is a very atypical example: that unprecedented international agreement was based on a strong scientific consensus, on a body of solid understanding (Rowland 1989). A much more typical situation is one of limited understanding, major uncertainties, and impossibility to offer a firm guidance for effective management. Reasons for this can be found in a combination of widespread objective research difficulties, limited utility of some preferred analytical techniques, and of inherent – but largely ignored – weaknesses of scientific inquiry.

ON PARTS AND WHOLES

There is, indeed, a difficulty about part and whole . . . namely, whether the part and the whole are one or more than one, and in what way they can be one or many, and, if they are more than one, in what way they are more than one.

Aristotle, *Physics*

Modern science emerged and has flourished as an assemblage of particularistic endeavours, a growing set of intensifying, narrowly-focused probes which have been bringing great intellectual satisfaction as well as rich practical rewards. But integration embracing interaction of complex, dynamic assemblages – what came to be called systems studies – is a surprisingly young enterprise. Smooth functioning of many biospheric interactions is indispensable for civilization's survival and to understand how the whole works is impossible by just focusing on disjointed parts. Key variables may be easy to identify but difficult to know accurately, and what should be included in order to represent properly the basic dynamics of the system is almost always beyond our reach.

I will reach for an outstanding example to the essentials of human survival. For more than a century it has been appreciated that life on the Earth depends critically on incessant transfers of nitrogen from the atmosphere to soils, plants, and animals and waters and then back to the air: maintenance of high rates of nitrogen fertilization is a key precondition for continuation of highly productive farming. Clearly, we should know a great deal about the nitrogen cycle – and we do. None of the grand biogeochemical cycles has been studied for so long, so extensively and in such a detail, and the sum of our knowledge is impressive (Bolin and Cook 1983; Smil 1985a; Stevenson 1986).

Fertilizing a field

And yet when a new spring comes around and an Iowa farmer is to inject ammonia into his corn field, or a Jiangsu farmer is to spread urea on his paddy, scientific knowledge proves to be a very uncertain guide. Laboratories will test soils and recommend the optimum nitrogen application rate but this advice does not have the weight of a physical law. Identical soil samples can come back from different laboratories with recommendations entailing several-fold differences in the total fertilizer cost and more than a twofold disparity in the total amount of nitrogen to be applied per hectare (Daigger 1974). That crop fertilization remains as much an art as a science is the inevitable consequence of the astonishing complexity of the nitrogen cycle.

The final product of this system's operation – grain yield in this example – cannot be predicted on the basis of even a very accurate knowledge of such key parts as nitrogen content available in the soil, the plant's known nitrogen requirements or the fertilization timing and rates. And while we have identified every important reservoir and every major flux of the element, such qualitative appreciation of individual parts does not add up to a confident understanding of the whole. A farmer wishing to know the nitrogen story of his field would not only have to set down a long equation describing the dynamics of this intricate system – but he would have then to quantify all the fluxes (Figure 1.1).

Nitrogen inputs to his field will come naturally from the atmosphere, as nitrates (A) and ammonia (B) in precipitation and in dry deposition and as ammonia (C) and nitrogen oxides (D) absorbed by plants and soil. If an Iowa corn crop is to follow the last year's soybeans, *Rhizobium* bacteria living symbiotically with the roots of those leguminous plants have fixed considerable amount of nitrogen (E) and a much smaller amount will be contributed by free-living nitrogen fixers (F); soybeans, beans or vetch could have preceded a Jiangsu rice crop, which may get more biotically-fixed nitrogen from *Azolla* in paddy water. Both fields will receive not only plenty of synthetic fertilizer (G), but also animal manures (H), above all from pigs.

More nitrogen was put into the soil by incorporation of crop residues from the previous harvest (I). Weathering (J), microbial decomposition of soil fauna proteins in the soil (K), and nitrification (L), a microbial oxidation of ammonia held in the soils to nitrates, are other important flows. Storage should account for nitrogen used by the crop (M), and by soil microorganisms (N), and for ammonia absorbed by the soil's clay minerals (O). Outputs should trace ammonia volatilization (P), denitrification (Q), that is, microbial reduction of nitrate to nitrogen, nitrogen losses from plant tops (R), leaching of nitrates into ground waters (S) and streams (T), organic (U) and inorganic (V) nitrogen losses in eroded soil, losses owing to microbial attack (W), insect feeding (X), harvesting losses (Y) and, of course, the harvest taken to the farm or storage (Z).

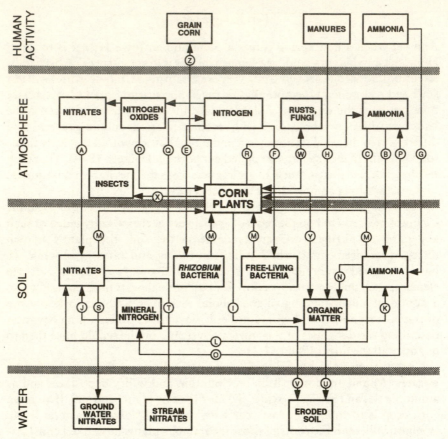

Figure 1.1 Intricacies of nitrogen cycling in a crop field preclude any accurate management of the nutrient's flows. Letters marking the flows are keyed to the discussion in the text.
Source: The cycle chart is modified from Smil (1990).

If the farmer would have numbers ready to be attached to this alphabet-full array of key variables he could fertilize with precision, with minimized losses and with the lowest costs. In practice, he knows exactly only one variable – the amount of synthetic fertilizer (G) he plans to apply to the field. How much will be removed in the crop he can only guess: drought can ruin his harvest – but a crop grown with optimum moisture could have used more fertilizer to produce a much higher yield (Figure 1.2).

Nitrogen in manures and crop residues (H, I) can be readily estimated – but with errors no smaller than 25–35 per cent. Estimating nitrogen in precipitation is much more uncertain but there are at least sparse measurements to provide some quantitative basis. Values on dry deposition and absorption (A–D) are almost universally lacking: we simply do not have even adequate techniques to measure these inputs properly.

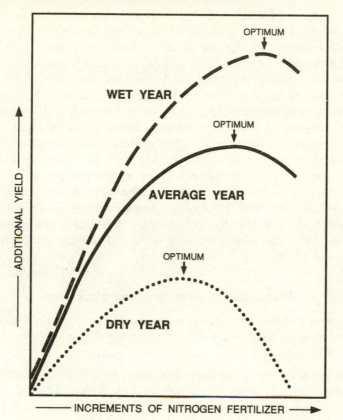

OPTIMUM

WET YEAR

OPTIMUM

AVERAGE YEAR

OPTIMUM

DRY YEAR

ADDITIONAL YIELD

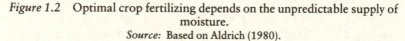

INCREMENTS OF NITROGEN FERTILIZER

Figure 1.2   Optimal crop fertilizing depends on the unpredictable supply of
moisture.
Source: Based on Aldrich (1980).

Fixation by symbiotic bacteria (E) and free-living organisms (F) is undoubtedly the most important natural input of nitrogen but we do not have accurate rate estimates even for leguminous crops which have been studied for decades in experimental plots: published extreme rates for most common leguminous species show five to sixfold differences (Dixon and Wheeler 1986). As for other input rates, all we can offer are just the right orders of magnitude and the same is true for all common nitrogen losses (P–X): again, there is no shortage of measured values but, as all of the processes are critically influenced by many site-specific conditions, adoption of typical values may result in large errors.

Even on this fairly small scale reliable understanding eludes us. I dwelled on this example because it deals with a mundane yet fundamental activity replicated many millions of times every year around the world. We may be sending probes out of our planetary system but it is beyond our capability to chart and to forecast accurately nitrogen fluxes of a single crop field and hence to optimize fertilization. But do not the current imperfect practices work? Indeed, but what portions of those fertilizers leak into waters and atmosphere where we do not want to have more nitrate or N_2O to contaminate wells, streams, lakes and bays, to acidify precipitation and to contribute to planetary warming and destruction of stratospheric ozone?

How much money – nitrogen fertilizer is almost always the single largest variable cost in modern farming – and labour is wasted, and how sustainable can the cropping be with large nitrogen losses in eroding soils and with declining reserves of the element in formerly nitrogen-rich ground? We may be maximizing the output, but we are very far from optimizing the performance of the whole system because we are not managing our agro-ecosystems in a sustainable manner, and because we are contributing to advancing environmental degradation. And so it really does not work – and even our best understanding assiduously translated into most careful agronomic practices allows for only a limited number of improvements (Runge *et al.* 1990).

Understanding plants and using black boxes

This inability to gain a satisfactory holistic grasp comes up repeatedly in the study of life and a single plant defeats us no less than a single field. Clifford Evans (1976) put it perfectly:

> In the broadest terms the main difficulties lie in the inaccessibility of the plant growing in its natural surroundings – physical inaccessibility because the great majority of methods of investigation involve gross interference with the plant, or with its environment, or both . . . and intellectual inaccessibility. The human mind has no intuitive understanding of higher plants: understanding . . . can only be reached by intellectual processes which attempt to follow through the consequences of particular observations and to integrate them with others.

Yet the overwhelming majority of those particular observations has come from studying plants in artificial environments, in enclosed growth chambers or in greenhouses. This eliminates the omnipresent natural interferences determining photosynthesis of free-growing plants. And even a perfect replication of natural environment would founder on the problem of time-scale: how to integrate, interpret and extrapolate brief observations to understand lives of plants spanning months, years and decades? Raup (1981a) suggested a new foundation for the study of plants which would not be based solely on their forms, processes and organization:

> Perhaps we should start with the inherited capacities of its species for adjustment to lethal disturbances that come from outside agents. The

frame would have a large element of randomness, but there would be no more randomness than the species have been coping with for a long time.

Physicists like to illustrate their difficulties with Heisenberg's famous uncertainty principle. In fixing definite position and velocity of a particle any conceivable experiment carries uncertainty and it is simply impossible to measure precisely both variables (Cassidy 1992). An ecologist could be hardly sympathetic. The physicists' lack of absolute certainty is replaced by a probabilistic expression: impossibility to pinpoint does not mean the loss of control and understanding. In contrast, an ecologist cannot replicate identical conditions in a plant or an ecosystem to pinpoint their behaviour. Field experiments with ecosystems are at best extremely difficult and very costly, and most often they are simply impossible. Unlike in physics or chemistry, most of the evidence in ecology is non-experimental and hence open to radical shifts in interpretation.

A generation of rapidly expanding ecological modelling has not been crowned with a confident outlook. Peters (1991) found that the predictions 'are often vague, inaccurate, qualitative, subjective and inconsequential. Modern ecology is too often only scholastic puzzle-solving'. Life's complexity accounts for a large part of these failures: as Edward Connor and Daniel Simberloff (1986) concluded, 'ecological nature is more complex and idiosyncratic than we have been willing to concede'. Plants remember everything. We are unable either to intuit this whole or to study it as a whole. We have to dissect to understand but, while our particular observations may reach amazing depths, they may be of very limited value in piecing together the grand picture. And during these dissections, even when working with caution, we can irretrievably destroy some key parts or relationships.

Once again Clifford Evans (1976): 'The biologist is always confronted by a player on the other side of the board. A false move, and the system on which the biologist thinks he is working turns into something else'. Emerson (1841) understood this perfectly: 'We can never surprise nature in a corner; never find the end of a thread; never tell where to set the first stone'. Understanding of social processes faces these obstacles on a still more complex plane. Herbertson's (1913) observations sum up this challenge:

> It is almost impossible to group precisely the ideas of a community into those which are the outcome of environmental contact, and those which are due to social inheritance . . . It is no doubt difficult for us, accustomed to these dissections, to understand that the living whole, while made up of parts with different structures and functions, is no longer the living whole when it is so dissected, but something dead and incomplete.

13

The crop-growing example also helps to illustrate the problem of parts and wholes when seen from the viewpoint of individual competence. Trying to understand behaviour of complex systems through an ever finer dissection of their parts is a universal working mode of modern science, a source of both astonishing discoveries – but also of massive ignorance spreading throughout our black-box civilization. Naturally, traditional farmers did not have any understanding of the physical or biochemical bases of their world but by observation, trial and error they reached a surprising degree of control over their environment.

They grew their crops in rotations we still find admirable, combined soybeans and rice, lentils and wheat or beans and corn in a single meal (to receive proper ratios of all essential amino acids from an overwhelmingly vegetarian diet), could build their houses, stoves and simple furniture, and the meadows and forests furnished them with a herbal pharmacopoeia which contained plenty of placebos but also many ingredients we still find irreplaceable. I do not want to idealize the hard life of traditional peasants – just to point out that even a small village could replicate all the modest but fairly efficacious essentials of that civilization.

Possibilities of average human life were greatly circumscribed, yet it could be seen as a whole, and open to risky, but decisive, action. Our positions are fundamentally different as the vast accumulated knowledge is almost infinitely splintered and we produce things, manage our affairs and lay out long-range plans by constant recourse to black boxes. A perfect illustration of these dealings is Klein's (1980) sci-fi story. As a group of scientists from Harvard's Physics Department 'pull back' a man present in the same building centuries from now, they are overjoyed that he is a professor not a janitor. But he can answer their questions on how numerous scientific breakthroughs were achieved and how the future works only with repeated 'that's not my field'.

The only realist in the time-breaking group is not surprised as he asks: 'How much do you think you'd know if you went back into the Dark Age? Could you tell them how to build an aeroplane? Or perform an appendectomy? Or make nylon? What good would you be?' Even the very people doing these things would be of little help as they, in turn, rely on numerous black-box ingredients and processes whose provision is in the hands of other specialists, who, in turn use other black boxes, circuitously *ad infinitum*.

Industrialization has done away with many existential strictures – but also with the relative simplicity of action. As the complexities of modern society have progressively reduced the effects of natural forces they have also limited the scope of individual action: as greater freedoms require greater control every change has cascading effects far beyond its initial realm. Jacques Ellul's (1964) lucid explications see the central problem of modern civilization in its dominance by *la technique*: it gives us unprecedented benefits and almost magical freedoms but in return we have to submit to its rules and strictures.

Modern man knows very little about the totality of these techniques on which he so completely depends, he just follows its dictates in everyday life: the more sophisticated the technique the harder the understanding of the underlying wholes and the more limited the true individual freedom. A large part of humanity is already irrelevant to production process and, although in so many ways it may be much freer than its ancestors, it certainly feels its slipping control of things. One may not agree with every detail of Ellul's arguments but it is impossible to deny the fragmentation of understanding accompanying advances of the technique.

Resulting reversal of human perceptions of the parts and the whole has profound effects on our understanding and management of environmental change. Our ancestors were awed by the intricate working of the creation whose mysteries they never hoped to penetrate – but individually they had an impressive degree of practical understanding and a surprising amount of effective control of their environment. In contrast, as a civilization, we are either in control of previously mysterious wholes (such as breeding cycles of many infectious diseases), have at least sufficient early warnings (by, for example, watching hurricane paths from satellites), or feel confident that we can understand them once we put in more research resources.

But individually the overwhelming majority of people, including the scientists, has no practical understanding of civilizational essentials and our controls rest on mechanical application of many black-box techniques. Our understanding resides in a myriad of pieces forming a web of planetary knowledge which Teilhard de Chardin (1966) labelled the noosphere. Obviously, it is much easier to use this large reservoir of particularized expertise as a base for even more specialized probes, a strategy responsible for rapid advances of scientific subdisciplines.

With unprecedented modes of environmental management – be they cost-driven or command-and-control adjustments on levels ranging from individual to global – we would be adding more levels of complexity. Underlying reasons for such changes, often disputed by the elites in charge, would be generally understood only in terms of simplest black-box explanations – but they would require profound transformations of established ways of living and they would have obviously enormous impacts on the whole society. But during the past generation we have witnessed an intellectual shift aimed at rejoining the splintered knowledge into greater wholes as quantitative modelling, greatly helped by rising powers and declining costs of computers, moved far beyond the traditional confines of engineering analyses to simulate such complex natural phenomena as the Earth's climate or the biospheric carbon cycle.

So explosive has been this reach that one can find few complex natural or anthropogenic phenomena that have not been modelled. These exercises are no longer mere intellectual quests: when the decisions-makers turn to scientists for advice in complex policy matters, computer modelling now has almost always a prominent, if not the key, role in formulating the recommendations: decisions shaping the future of the industrial civilization originate increasingly with these new oracles. According to the ancient Greeks, the centre of their flat and circular world was either Mount Olympus, the abode of the gods – or Delphi, the seat of the most respected oracle. Industrial civilization appears to have little left to do with gods. A possibility that we might be shifting the centre of our intellectual world into the realm of computer modelling makes me very uncomfortable. A closer look at the limits of those alluring exercises will explain why.

LIMITS OF MODELLING

And the Colleges of Cartographers set up a Map of the Empire which had the size of the Empire itself and coincided with it point by point . . . Succeeding generations understood that this Widespread Map was Useless, and not without Impiety they abandoned it to the Inclemencies of the Sun and of the Winters.

Suarez Miranda, *Viajes de Varones Prudentes* (1658)

Computer modellers must empathize with cartographers in Miranda's imaginary realm. Their predicaments are insolvable. Simple, large-scale maps will omit most of the real world's critical features and their heuristic and predictive powers will be greatly limited: often they will offer nothing but a restatement of well-appreciated knowledge in a different mode. Yet offering perfect, point by point, replications of complex, spatially extensive realities would presuppose such a thorough understanding of structures and processes that there would be no need for setting up these duplicates which may end up as the Widespread Map.

Modelling is undoubtedly a quintessential, and inestimably useful, product of human intelligence: it allows us to understand and to manage complex realities through simpler and cheaper (often also safer) thought processes, and computers elevated the art to new levels of complexity. But modelling is also always an admission of imperfect understanding. Alfred Korzybski's (1933) reminder is hardly superfluous: 'A map is not the territory it represents, but, if correct, it has a *similar structure* to the territory, which accounts for its usefulness . . . If the structure is not similar, then the traveller is led astray . . .'

Not surprisingly, structure of many models dealing with the biospheric change is not yet sufficiently similar to the complexities of systems subsuming multivariate interactions of natural and anthropogenic processes. This obvious reality does not discredit a use of inadequate models – but it requires to pay close attention to their limits. For any experienced modeller this is quite clear, but mistaking imperfect maps for real landscapes has been a common enough lapse: results of modelling exercises can come

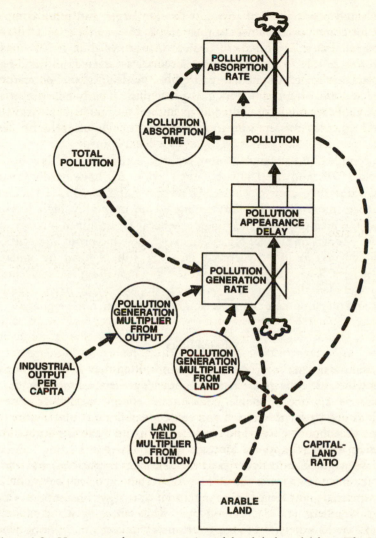

Figure 1.3 How not to forecast: a section of the global model from *The Limits to Growth* which used excessively large, and fundamentally meaningless, aggregates such as 'pollution', linked by no less abstract generation rates and multipliers.
Source: Meadows *et al.* (1972) *The Limits of Growth.*

out with plenty of attached qualifiers, but such information is the first to be lost when the conclusions are concentrated into executive summaries or squeezed into news columns. And, of course, there are always some modellers who make unrealistic claims for their creations: their reasons may range from understandable enthusiasm to a dubious pursuit of large grants.

The error of mistaking maps for realities has been strengthened with the diffusion of computer modelling. For modellers the machines opened up unprecedented possibilities of toying with ever larger and more complex systems, for common consumers of their efforts they lent a misplaced aura of greater reliability. *The Limits to Growth* (Meadows *et al.* 1972) – based on Jay Forrester's (1971) business dynamics models – started this trend right on the planetary scale. An untenably simplistic model of global population, economy and environment (Figure 1.3) predicting an early collapse of the industrial civilization was presented by its authors as a realistic portrayal of the world and this misrepresentation was uncritically accepted by news media.

More complex and somewhat more realistic computer models followed during the 1970s and the 1980s but their print-outs have hardly greater practical value than their celebrated predecessor (Hafele 1981; Hickman 1983). Their most obvious shortcoming is the inevitably high level of aggregation which results in meaningless generalizations or in incongruous groupings. Models on smaller scales can avoid these ludicrous aggregations but even when carefully structured they are still plagued by inherent incompleteness. Reviewing the state-of-the-art in modelling reality, Denning (1990) concluded that 'experts themselves do not work from complete theories, and much of their expertise cannot be articulated in language'.

Missing links

These weaknesses are especially important in modelling those human activities which have the greatest effect on the global environment. A perfect illustration is an unpardonable omission in energy modelling. Energy conversions are by far the largest source of air pollutants and a principal source of greenhouse gases, and in order to avoid health, material and ecosystemic damage and to prevent potentially-destabilizing climatic changes we would like to understand the prospects of national and global energy use. Since 1973 (after the OPEC's first crude oil price rise) modellers have constructed many long-range simulations of energy consumption – but, as Alvin Weinberg (1978) pointed out, while there is no doubt that historically we have been led to use more energy to save time, among scores of post-1973 energy models not a single one took into account this essential trade-off between time and energy.

Instead, they focused on achievement of politically-appealing but economically-dubious self-sufficiency, on ill-defined diversification of supply (with ludicrously-exaggerated assumptions about the penetration rates of new techniques) or on a rigid pursuit of maximized thermodynamic efficiency where everything is subservient to the quest for better conversions (a goal which may make little economic or environmental sense). This crucial omission can be remedied: time-saving performance can be incorporated into more realistic simulations but there are many other essentials left out of all models of energy supply and demand, as well as from their more general

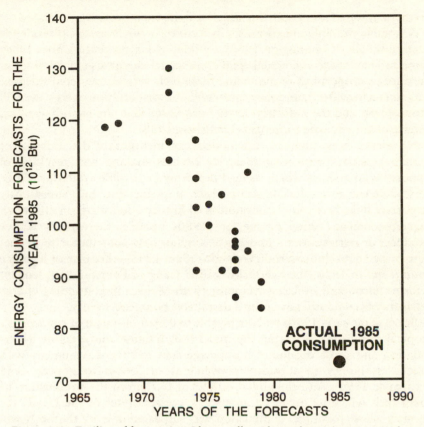

Figure 1.4 Futility of forecasting. I have collected two dozen American mid- to short-term energy forecasts of the country's total primary energy consumption in 1985. Expectedly, the closer they were made to the target year, the more realistic they tended to be, but even those looking ahead less than a decade were, on the average, about 25 per cent too high.

analogies designed to forecast the course of national economies or even the trends of global developments.

Unpredictable personal decisions – ranging from a sudden abdication of a shah to a course taken by a small group of men in a white stucco building on Pennsylvania Avenue to an inexplicable miscalculation of an aggressive dictator, events whose eventual effects the protagonists themselves could not imagine – have a profound long-term impact on the environment as they can lead to rapid shifts in energy prices or, as seen in Saddam Hussein's destruction of Kuwaiti oilfields, directly to enormous environmental damage. After two generations of forecasting studies we can point to a number of utterly wrong forecasts. Even relatively short-term forecasts almost invariably miss the eventual reality by wide margins (Figure 1.4). But much more interesting is a smaller category of studies whose overall

19

predictions were fairly good, but which still missed the essential changes.

A notable example are forecasts in *Resources for Freedom* (President's Materials Policy Commission 1952). Looking a quarter century ahead they came rather close to the actual total commercial energy use in 1975 – but they got its proportions of individual fossil fuels very wrong (coal's demise was so much swifter!), and they forecast that transportation energy use will grow slower and the industrial energy use faster than the overall average, reverse of the actual development (Landsberg 1988).

No matter how complex, no models can incorporate the universe of human passions permeating everyday decision-making and resulting in tomorrow's outcomes which seemed decidedly impossible yesterday. Three very different examples illustrate these unpredictable but enormously important links between discontinuities of history and long-term effects on the environment. China of the late 1980s became the world's largest producer of refrigerators – and hence a major new contributor to ozone-destroying chlorofluorocarbon emissions (Smil 1992) – because of a palace *coup d'état* in 1976 (the overthrow of the Gang of Four) and subsequent reforms introduced by Deng Xiaoping: with Maoists in continuing charge refrigerators could still have been a despicable bourgeois luxury.

Brazil of the early 1990s will not be able to destroy as much of Amazonia's tropical rain forest as during the mid-1980s because the years of hyper-inflation and gross economic mismanagement left the government with much less money to subsidize conversions of forests to grazing land (Fearnside 1990). And the global output of CO_2 from the combustion of fossil fuels will slow down because the successor states of the USSR, the country whose fossil fuel combustion was surpassed only by the USA, will first produce less energy because of economic disarray, and eventually they will need less of it because of efficiency improvements. None of these considerations entered any global warming models – yet their cumulative impact will be clearly enormous.

Obviously, using inappropriate models to deal with complex realities can retard rather than advance effective solutions. But the problem is deeper: as Rothenberg (1989) pointed out,

> there are even cases where using any model at all is inappropriate. For example, relating to someone in terms of a model precludes the comprehension and appreciation of the richness and unpredictability that distinguish living beings from rocks and tractors.

Replace 'someone' by 'anthropogenic environmental change' – and you are touching the very core of difficulties and inadequacies faced by modellers trying to represent the impact of those transformations on modern civilization. No model can be good enough to enlighten us about our future collective capacity for evolutionary adaptation.

But is not all this trivial? Are not the unpredictabilities of individual human actions and stunning discontinuities of history obvious hallmarks of civilizational evolution? Indeed, but I dwelt on these points precisely because they are repeatedly ignored, ignored in ways which are both costly and embarrassing. Trends and concerns of the day are almost always casting long shadows over the future, and smooth functions of expected developments, linear or exponential, are seen to continue far ahead. Unpredictable discontinuities cannot be, by definition, included – thus excluding any chance of a truly realistic assessment.

What is much less explicable is that the reigning trends and concerns of the day have such an inordinate influence even in the case of cyclical phenomena which are characteristic of a large number of environmental changes. Of course, where complex socio-economic systems are concerned, the existence of cyclicity cannot be equated with easy predictability – frequencies and amplitudes may be far from highly regular – but the inevitability of trend reversals is obvious. Incredibly, during the past generation many modellers have repeatedly ignored these realities, transfixed by the impossible trends of permanent growth or decline.

The record of economic modellers has been most unimpressive, in no small part a result of the profession's virtual abandonment of reality. Wassily Leontief (1982) exposed the ruling vogue in contemporary economics by noting that 'page after page of professional economic journals are filled with mathematical formulas leading the reader from sets of more or less plausible but entirely arbitrary assumptions to precisely stated but irrelevant theoretical conclusions'. Watt (1977) looked at the dispiriting forecasting record and found perhaps the fundamental problem in gross underestimation of the magnitude of the task involved in building such interactive models: it is surprisingly hard to put together all necessary submodels which one might expect to be readily available from specialists.

Watt suggested two general reasons for this. First, the overwhelming concentration of academic research on clearly defined problems rather than on complex refractory topics with fuzzy boundaries (parts and wholes again). Obviously, this offers the easiest road for career advances but it makes for a near-permanent choke on widespread system orientation in sciences. Second, a shortcoming already stressed just a few paragraphs ago: a strong preference of most scientists to study phenomena within an implied steady-state environment, assuming the futures of continuous growth or decline and ignoring many non-linear forces and discontinuities.

Furthermore, Watt was frustrated by the impossibility of conveying the findings in a complex, scientific manner as the decision-makers in the western setting treat the problems largely in accord with linear thinking of constituencies they are trying to satisfy in order to assure their political survival. He urged great efforts in communicating the results in language which the legislators will understand. Watt was after an understandable and desirable goal – an effective diffusion of complex scientific understanding –

but with the public increasingly unable to understand even the rudiments of our advances (the black-box syndrome) this is clearly a losing proposition. And I would not hesitate to put a large share of blame for complex quantitative models on the modellers themselves.

Their work has inherent limitations and loud and repeated acknowledgment of this fact would serve them well – but when this is done it is usually in the fine print of an appendix. Modellers' common transgressions are clustered at the opposite ends of the utility spectrum. On the one hand the non-specialists in quantitative analysis and computer science, driven by a peer pressure into complex modelling, have often a very poor understanding of the capabilities of the whole exercise: exaggerated expectations have been quite common. Many scientists succumb readily to the lure of print-outs 'revealing' the future. Building a big model is often intellectually challenging, running it is exciting, glimpsing what they believe to be a possible future is a matter of a classy privilege. There seems to be also a distinct cultural preoccupation betraying the American can-do, fix-it attitude to complexities.

But what about the strictly natural phenomena? Are there not many examples of highly successful simulations of complex natural processes which can be described by a finite number of reactions or equations and precisely quantified by repeated measurements? And are not these endeavours getting steadily more rewarding? Yes in all cases – but only for certain classes of problems. Strict reproducibility of relationships and processes and homogeneity of structures in the inanimate world form a splendid base for the great predictive powers of the physical sciences. But even in this calculable world computer models are frequently defeated by the sheer complexity of the task, by the multitudes of non-linear ties which make any simulation accurate enough to be used with confidence in decision-making, not just as a heuristic tool, elusive.

Computers, clouds and complexities

Certainly the best example in this intractable category is the modelling of the atmosphere in general, and of any future climatic changes in particular. Even the best current climatic models still have only a rough spatial resolution: their grids are still too coarse and the vertical changes in the principal atmospheric variables are modelled on separate levels. Grids of one degree of latitude by one degree of longitude and forty vertical levels would bring us much closer to the reality: it would have up to 100 times more points than the existing models, but even these exercises are running into computing problems (Mahlman 1989).

Computing power keeps on increasing, but its growth is far from being matched by deeper understanding of ocean circulation, cloud processes and atmospheric chemical interactions: hence even greatly expanded computing capacities would not result in highly reliable simulations. Temperature rise determining the severity and the consequences of the anticipated global warming is the most critical variable, but its predictions come from models which can replicate some of the grand-scale features of global climate but whose current performance is incapable of reproducing the intricacies and multiple feedbacks of natural phenomena.

And although some of these models may be in relatively good agreement as far as the mean global temperature increase is concerned, the overall range (as little as 1.5° and as much as 4.5°C with doubled pre-industrial CO_2 levels) remains still too great to be of any use for practical decision-making. Differences in regional prognoses are much greater, precluding any meaningful use of these simulations in forecasts of local environmental effects. Schlesinger and Mitchell (1987) had to conclude that existing models were frequently of dubious merit because they were physically incomplete or included major errors. Their comparisons showed even qualitative differences. This means that not all of these exercises can be correct, and all of them could be wrong. Treatment of clouds and oceans has been particularly weak.

General circulation models can represent essentials of global atmospheric physics but the iterative calculations are done at such widely-spaced grid points that it is impossible to treat cloudiness in a realistic manner. Even the highest-resolution models use grids with horizontal dimensions of about 400 km, an equivalent of New England and New York combined or half of Norway, and less than 2 km thick vertically (Stone 1992). And yet the clouds are key determinants of planetary radiation balance (Ramanathan *et al.* 1989). A National Aeronautics and Space Administration (NASA) multi-satellite experiment showed that the global 'greenhouse' effect of clouds is about 30 W/m^2; in comparison, doubling of pre-industrial CO_2 concentration would add only about 4 W/m^2 (Lindzen 1990). And because the warming effect of higher atmospheric CO_2 levels increases only logarithmically, future tropospheric CO_2 levels would have to be more than two orders of magnitude higher than today to produce a comparable effect.

In contrast, the clouds' cooling effect is close to –47 W/m^2, resulting in the net effect of about –16 W/m^2. Without this significant cloud-cooling effect the planet would be considerably (perhaps by as much as 10–15 K) warmer. Not surprisingly, Webster and Stephens (1984) estimate that a mere 10 per cent increase in the cloudiness of the low troposphere would compensate for the radiative effects of doubled atmospheric CO_2. Of course, these facts do not guarantee that the clouds will continue to have such a strong cooling effect in the future. Deficiencies in modelling oceans are no less fundamental. Models assume continuation of their poorly understood present behaviour – but that will most likely also change with changing climate. For example, shifts in oceanic circulation may create major CO_2 sinks and alleviate the tropospheric accumulation of the gas (Takahashi 1987).

The smallest forecasts of planetary warming during the next two generations could be largely hidden within overall climatic noise, while the highest calculated response would have a profound effect on snow and ice cover and on the latitudinal climatic boundaries. Where does this leave us? The only reasonable generalization is about the direction of the change: increased concentrations of greenhouse gases will most likely increase surface temperature. But this response was described by John Tyndall (1861) and detailed by Svante Arrhenius (1896) generations ago without computer models and six-figure research expenditures.

And in spite of the continuous growth of computing power and initiation of ambitious national programmes to get much better models, we may not be able to tighten the existing uncertain forecast of 1.5°–4.5°C with doubled pre-industrial CO_2 levels during the 1990s – unless the calculating speeds go up by many orders of magnitude. Because important variations in moisture convection proceed on scales between 100–1000 m we would need grids 1000 times finer than those with 400 km sides, iterations should be done about 1000 more frequently, and vertical resolution would have to improve. These requirements would demand computing speeds about 10 billion times faster than in the early 1990s (Stone 1992).

Initializing such detailed models would also require massive observation effort gathering a huge volume of missing data, and much of new thinking would have to be infused into the modelling. The latter ingredient is probably decisive. J. D. Mahlman, director of the National Oceanic and Atmospheric Administration's Geophysical Fluid Dynamics Laboratory, concludes that the problem 'is so quantitatively difficult that throwing money around is not going to solve it by 1999 or whenever' (Kerr 1990a). But even if the modellers would converge on an identical value we should not take it as an unchallengable result because their exercises are largely one-dimensional, their calculations supplying the answers to questions about what would happen as just one variable changes.

Staying with the global warming example, future tropospheric temperatures will not be only a function of rising emissions of principal greenhouse gases, but will be also influenced by changed stratospheric chemistry, by altered oceanic circulation, by higher concentrations of sulphates and nitrates generated by growing fossil fuel combustion, by extensive changes in the Earth's albedo and by modification of ecosystems. Some of these links will have amplifying effects, others will have damping impacts: we simply do not know what the net result may be two or three generations ahead.

Change, in nature or in society, is never unidimensional: it always comes about from interplays – additive or multiplicative, synergistic or antagonistic. Dynamics of the process may make irrelevant even a successful identification of the key initial causes: after all, what would be the point of finding the match which lit the fire? There is an even more fundamental limit to the usefulness of quantitative models even in apparently well-definable physical variables: the possibility that the underlying phenomena may not be encompassed by mathematical expressions owing to the fact that in non-linear systems an infinitesimal change in initial conditions can result in entirely different long-term outcomes.

Atmospheric behaviour offers perfect examples of these limitations. Because the basic equations describing atmospheric behaviour are non-linear, Edward Lorenz (1979) found in his pioneering models that numerical inputs rounded to three rather than to six decimal places produced very different forecasts. As it is impossible to establish all the necessary inputs with an identical degree of accuracy, rounding, actual or by default, is always present and models may correspond very poorly even to the next week's reality. Ultimately, it means that accurate long-range weather forecasting is most likely impossible.

Other natural phenomena display similarly chaotic behaviour: they do not evolve into a steady state, are always changing in a non-linear and unpredictable fashion, but remain within certain limits and their behaviour is marked by spells of stability (Gleick 1987). Poincare (1903) captured this reality long before its popularization by the modern chaos theory:

> It may happen that small differences in the initial conditions produce very great ones in the final phenomena. A small error in the former will produce an enormous error in the latter. Prediction becomes impossible, and we have the fortuitous phenomenon.

Still, the homogeneity of many structures and reproducibility of most processes governing the inanimate world form a solid base for gradual advances.

In contrast, the variability of individuals and uniqueness of events is the essence of living. Individuals are unsubstitutable, phenomena are unrepeatable in ways which would produce identical outcomes. Slobodkin (1981) has an excellent analogy which calls for imagining a number of independent universes, none of them more important than any other, so that the physicists would have to follow a procedure of building 'the theory on superficial information about all universes rather than detailed information about any one of them'. Physics would be then in a situation somewhat akin to the study of life complexes as ecosystems are likely to differ in the same way, and probably to an even greater extent than the hypothetical separate universes. In Cohen's (1971) summary, 'physics-envy is the curse of biology'.

But all modellers must, sooner or later, put more of their personalities into the models. Quade (1970) summed up this ultimate constraint perfectly:

> The point is that every quantitative analysis, no matter how innocuous it appears, eventually passes into an area where pure analysis fails, and

subjective judgement enters ... judgement and intuition permeate every aspect of analysis: in limiting its extent, in deciding what hypotheses and approaches are likely to be more fruitful, in determining what the 'facts' are and what numerical values to use, and in finding the logical sequence of steps from assumption conditions.

Finally, an important common limitation introduced by the modeller. Abundant evidence shows complex systems behaving in a wide variety of ways. If one wants to impose order then assorted fluctuations or cycles would fit best. Not so with a disproportionately large share of computer models of environmental changes and socio-economic developments. They carry warnings of resource exhaustion, irreversible degradation, runaway growth, structural collapses, pervasive feelings of pre-ordained doom. They display one of the frequent failings of science: pushing preconceived ideas disguised as objective evaluations.

FAILINGS OF SCIENCE

This science is much closer to myth than a scientific philosophy is prepared to admit. It is one of the many forms of thought that have been developed by man, and not necessarily the best. It is conspicuous, noisy, and impudent, but it is inherently superior only for those who have already decided in favour of certain ideology, or who have accepted it without ever having examined its advantages and its limits.

Paul Feyerabend, *Against Method* (1975)

In his iconoclastic critique of scientific understanding Paul Feyerabend (1975) noted that even 'bold and revolutionary thinkers bow to the judgement of science ... Even for them science is a neutral structure containing positive knowledge that is independent of culture, ideology, prejudice'. This is, of course, a myth. Science is as much a human creation as religion and legendary tales, with its basic beliefs protected by taboo-like reactions, with its unwillingness to tolerate theoretical pluralism, with its prejudices and falsifications, with its irrational attachments and alliances, and with its changing orthodoxies. Its practitioners have too much at stake to be interested in nothing but intellectual challenges of their efforts.

These considerations must be kept in mind when appraising our knowledge of environmental change and all the advice for action which the scientists rain on the society. They are very helpful in understanding one of the most notable features of contemporary environmental studies: overwhelming gloominess in assessing the planetary future. One way to understand this is to realize that in the western industrial civilization science has come to perform several key functions previously reserved for religion. Perhaps, as Zelinsky (1975) noted, 'the worship of science is the only authentic major religion of the twentieth century'.

Above all, science explains the world according to its laws and it does not suffer questioning of its canons lightly. And from its complex hierarchy of priesthood (from BS to FRS, from lecturers to Nobel Prize laureates) there must rise scholars not content with daily copying of missals and *a capella* chantings, that is, writing yet another paper for the *Journal of XYZ Association* and delivering its mutant at a conference where everybody wears a name and a glass during recess.

Prophets as catastrophists

In a time-honoured practice prophets coming from the long line of Graeco-Judaeo-Christian tradition must be bringing bad news. In fact, the worse the news, the greater the prophecy. To compete with the best known one is difficult: 'And death and hell were cast into the lake of fire. This is the second death. And whosoever was not found written in the book of life was cast into the lake of fire' (*Revelation* XX:14–15). Scientific prophets do not promise the second death but the first one they wholesale freely. Obviously, theirs is not the whole science story – but their opinions get invariably much more attention than reserved assessments of scientists trying to cope honestly with the biospheric complexity.

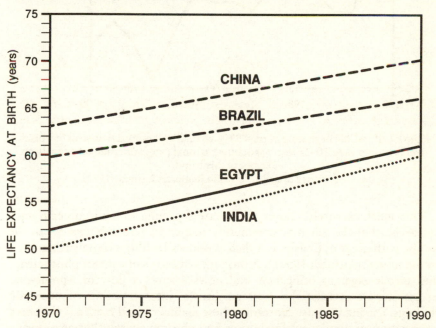

Figure 1.5   During the past generation life expectancies increased appreciably in every poor populous nation. China's record has been especially impressive: now only a few years separate the country from averages in European nations.
Source: Plotted from data in the World Bank (1991).

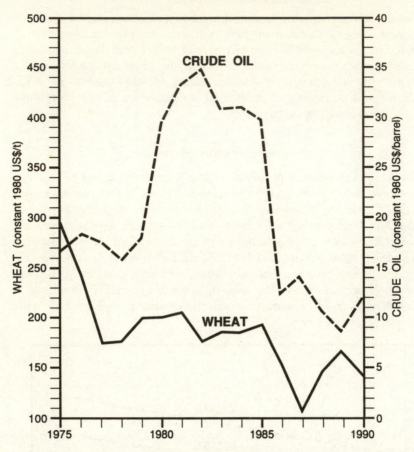

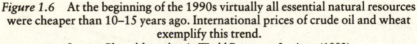

Figure 1.6 At the beginning of the 1990s virtually all essential natural resources were cheaper than 10–15 years ago. International prices of crude oil and wheat exemplify this trend.
Source: Plotted from data in World Resources Institute (1992).

This must be surely one of the greatest peculiarities of contemporary research: 'that the scientific community relates its research agendas to the public with great optimism, but then produces terribly pessimistic analyses of the future' (Ausubel 1991a). A thriving symbiosis of catastrophist science and media eager to bring new bad news assures a flow of Apocalyptic prognoses, or at least the most pessimistic interpretations of uncertain findings. During the past generation these claims ranged from a widespread expectation of permanent food crises and chronic energy shortages during the 1970s (Brown 1976) to an inevitability of catastrophic climatic change and sea-level rise (Schneider 1989).

Yet both the realities and the best available assessments at the beginning

28

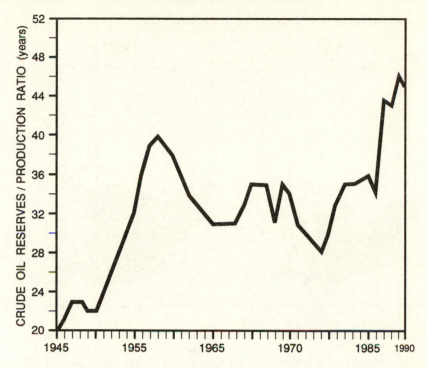

Figure 1.7 No imminent exhaustion: world-wide reserve/production ratio for crude oil stood higher in the early 1990s than at any time since *The Oil and Gas Journal* started to publish its annual review of reserves and production.

of the 1990s were very different. China, India and Indonesia, containing about one-half of the poor world's population, were self-sufficient in grain and (with the exception of countries suffering from civil war) average life expectancies rose everywhere (Figure 1.5). International wheat and crude oil prices were lower than a generation ago (Figure 1.6), and the world-wide reserve-production ratio for crude oil was higher than during any year since 1945 (Figure 1.7). Satellite monitoring showed no sign of global warming during the 1980s (Spencer and Christy 1990), and the most detailed appraisal of sea-rise prospects did not endorse scenarios of high flooding (*Ad hoc* Committee on the Relationship between Land Ice and Sea Level 1985).

But realistic assessments receive much less media attention than catastrophic news, and careful refutations of exaggerated claims are even harder to find. Two outstanding recent examples of the latter phenomenon include the matters of the expanding Sahara and the disappearing frogs, processes widely seen as important indicators of environmental degradation. Declining frog numbers received special attention because frogs were seen as sensitive indicators of the declining habitability of the planet.

An older perception of the expanding Sahara has received much attention

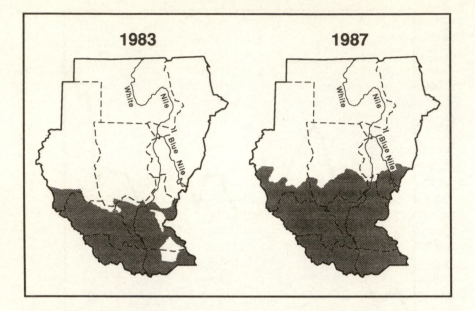

Figure 1.8 Sequential satellite images of Sudan show how the border between vegetation and desert in the Sahelian zone fluctuates with rains. Contractions (shown here during a dry year of 1983) and expansions (during a wet year of 1987), rather than any southward march, describe changes in the region's plant cover.
Source: This simplified extent of plant coverage is derived from satellite images in Hellden (1991).

ever since the great Sahelian drought of the early 1970s, and the southward march of the desert has been a widely-cited illustration of environmental degradation. Eckholm and Brown (1977) had 'little doubt ... that the desert's edge is gradually shifting southward'; the area claimed by this march was given as an equivalent of New York state every decade (Smith 1986), and it was claimed that the desert's front line was advancing in Sudan at a rate of 90–100 km a year (Suliman 1988). At least the last claim should have made any numerate reader suspicious: such an advance would have desertified all of Sudan in less than one generation!

But careful studies show no evidence of this southward march. Satellite monitoring of long-term changes records extensive natural fluctuations, with the Sahara expanding or contracting at a rate of over 600,000 km^2 a year (Tucker *et al.* 1990). Detailed studies of the Sudanese situation have not verified any creation of long-lasting desert-like conditions, no major shifts in the northern cultivation limit, no major changes in vegetation cover and productivity and no major sand-dune formation (Hellden 1991). Temporary changes of crop yields and green phytomass were caused by severe droughts – and were followed by a significant recovery (Figure 1.8). Have these finding been featured in 'green' literature? Similarly, ominous declines of frogs, claimed to be a world-wide phenomenon by 1990, may have a much better explanation in random fluctuations brought by drought (Pechmann *et al.* 1990).

Can this worst-case bias be explained only because of the Apocalyptic predilection of scientific prophets? I believe not. Often their frightening conclusions result from problems and processes discussed earlier in this chapter: from misunderstanding the relationships between parts and the whole, from ignoring cyclicity and underestimating the importance of discontinuity, and from naïvely mistaking today's dubious models for tomorrow's realities. Less charitably, this Apocalyptic tilt may be an obvious manifestation of self-interest. Catastrophism generates publicity, and to believe that most of the scientists shun the front-page, keynote-speaker, on-camera exposure would be to deny their humanity. Perhaps even more importantly, possibilities of perilous futures can lead to major grants and conferences and to the building of bigger computer models, all the roads to greater scientific glory.

Researchers are sensitive to questions about their motives, but I strongly agree with Ausubel (1991a) that 'it is inconsistent for us to question the self-interest of the military-industrial complex, the medical establishment and other groups, but not our own'. Tendency to lean toward pessimistic conclusions or advocacy of the worst-case scenarios may have yet another important motive. Big problems may need big solutions which will require extensive government involvement. Consequently, there would be large, and growing, bureaucracies to direct, and long-range plans to lay out. Many scientists have a deep-seated preference for this kind of social or environmental engineering based on their designs.

And as for the Apocalyptically inclined computer modellers, James Lovelock (1979) has perhaps the best explanation, a combination of one-dimensional lives and gambling-like addiction:

> Just as a man who experiences sensory deprivation has been shown to suffer from hallucinations, it may be that the model builders who live in cities are prone to make nightmares rather than realities. No one who has experienced the intense involvement of computer modelling would deny that the temptation exists to use any data input that will enable one to continue playing what is perhaps the ultimate game of solitaire.

Encouraging outcomes would clearly discourage further modelling, while alarmist scenarios require more work to detail the looming catastrophes.

Deeper weaknesses

Failings of science have much deeper reasons than the fallibilities of its practitioners. Every one of its principal, and seemingly self-evident,

31

methodological axioms – causality, eventual solubility of all problems and achievement of perfect understanding, and universal validity of its findings – is questionable. Randomness and indeterminacy of many physical and mental processes militate against simplistic causal explanations, limits to both observability and comprehensibility of complex systems preclude the ultimate understanding, and the evolutionary brevity of scientific endeavour and its inherent weaknesses negate any claims for universal validity.

In Zelinsky's (1975) merciless phrasing, the latter aspiration is

> an extreme case of intellectual parochialism or scientific *hubris* . . . How can one posit universal validity for conclusions extracted by means of a provisional system of logic from a biased set of data taken by less than infallible observers using imperfect instruments over a period of at most a few centuries . . . ?

And as far as our quantitative knowledge of the Earth's environment is concerned, the period extends merely over a few decades! But even if methodologically sounder foundations would eliminate catastrophist biases, and even if the possession of much deeper understanding would greatly reduce the challenge of uncertainty and the resulting dilemmas of un-decidability, scientific interpretation of complex realities would still face fundamental pitfalls of hierarchization, order and addition (Jacquard 1985).

As soon as our knowledge of objects or phenomena reaches the level where they cannot be adequately characterized by a single parameter, we cannot place them in a hierarchical order. Of course, environmental resources and processes can be understood only through sets of irreducible parameters, a reality excluding any meaningful ranking.

> When we are comparing one number to another, nonequality implies that one is greater than the other; when we are comparing sets, it implies only that they are different. This is not a plea motivated by moralistic considerations; it is a statement of logical fact.
>
> (Jacquard 1985)

And it precludes any meaningful ranking of endangered species to be saved, any 'correct' allocation of limited resources to reduce different forms of pollution, or any choice of 'best' divisions of national responsibilities in efforts requiring global co-operation.

Science's propensity for setting up hierarchical structures reveals also its dominant bias for order, organization and harmony. In environmental management this preference is repeatedly seen in attempts to protect natural ecosystems as paragons of harmonious existence, or in recent phobias concerning global climatic change. But preservation of the biospheric *status quo* is fundamentally anti-evolutionary and maintaining that any anthropogenic changes are at best undesirable and, more likely, disastrous triumphs of disorder above order is to misread the evolutionary record. Evolution of the Earth's life was driven precisely by major cosmic (orbital irregularities, collisions with celestial bodies, changes in solar luminosity), geogenic (plate tectonics and associated volcanism) and biogenic (interspecific competition and adaptation) departures from temporary equilibria.

During the past few centuries the scope of human actions resulted in some unprecedented rates of change (perhaps most notably decline of forests, disappearance of tropical species and increases of trace atmospheric components). Still, this does not mean that overriding goals of environmental management should be simply conservation of existing ecosystems and preservation of pre-industrial biospheric fluxes and concentrations. But science is generally a very poor guide when trying to establish what degree of 'disorder' – departure from 'natural' arrangements, levels and flows – we should be able to accept or at least to tolerate. Basic questions of global environmental management – what degree of rain forest destruction we should tolerate, or what level of planetary warming would be acceptable – cannot be decided by science alone.

Pitfalls of addition are rooted in the early usefulness of the operation and in its enormous practical utility: we add even before we can define numbers, and we repeatedly solve complicated problems by breaking down the observable reality into quantifiable components which can be added together. But too many biospheric realities cannot be deconstructed into additive segments, and their understanding requires a concurrent grasp of constituent complexities. This cannot be done without vastly more effective contributions by social sciences, but their performance has been pitiful. Their failures are readily exemplified by the near-total failure of economics to explain the real world, by our inability to forecast population growth, and by a notably skimpy attention to such fundamental questions of human existence as the nature and causes of social change and the rise and fall of civilizations.

Indeed, in social inquiries it is difficult 'to identify a single new theoretical concept widely accepted as both scientifically valid (using the criteria of the natural sciences) and humanly important' (Zelinsky 1975). In spite of these pitfalls and shortcomings, modern science has accumulated a substantial record of practical achievements which has helped to entrench the belief that it offers superior guidance because of its inherent realism, its commitment, in Ian Hacking's (1981) words, 'to find out about one real world'. But this record has been marred by repeated loss of objectivity resulting in lapses into what Rousseau (1992) calls pathological science. Many scientists clearly have no desire at all to find out about one real world, while others are preaching, much as the disputatious servants of the past religious clashes, according to their particularistic, parochial biases – and frequently ask society to pay for them.

And many answers are missing simply because nobody has asked the proper questions. Most of the problems in science are selected by the scientists themselves and the questions are formulated in ways making scientific solutions feasible. The same is true about the research requested by governments: general demands are first broken down to manageable components by assorted advisory committees and expert panels. The self-amplifying prestige of science works on all levels of scientific inquiry and the problems offering little promise of solutions will go largely unresearched.

The sense of proportion is critical. Of course, science has served us well, and a deeper scientific understanding is urgently needed to deal with the enormous challenges of environmental change: it is unimaginable that we could follow a different path. But this does not mean, as argued by most of the philosophers of science (the 'realists') that science is our best exemplification of rationality, our best guide for the future. In contrast, for the 'relativists' science is just one of many social constructs with no claim for inherent superiority. I am in great intellectual sympathy with the latter claim – but I am too much of a product, and a producer, of a scientific culture not to claim a good deal of uniqueness for western science, not to take pride in its achievements, and not to treasure the possibility of contributing a few small stones to its vast mosaic.

With the realists I admit readily that this science has been successful – but as a relativist I know that success is an amorphous term, that outcomes are always full of ambiguities and contradictions. Larry Laudan's (1984) attempt to synthesize the two stands coincides with my feelings:

> The methods of science are not necessarily the best possible methods of inquiry (for how could we conceivably show that?), nor are the theories they pick out likely to be completely reliable. But we lose nothing by conceding that the methods of science are imperfect and that the theories of science are probably false. Even in this less-than-perfect state, we have an instrument of inquiry which is arguably a better device for picking out reliable theories than any other instrument we have yet devised for that purpose.

But this qualified utilitarian success does not mean that scientific *Weltanschauung* offers the only valuable insight into human condition and the best guide for future action. I very much agree with Greene's (1989) perspective that in spite of its enormous breadth of concerns science seeks only a very limited kind of intelligibility. Its preoccupation is with functional relationship among phenomena, while both our philosophical and religious visions of life try to find an intelligibility that makes sense of the totality of human experience. This includes knowledge of good and evil which is at the core of our transcendence and humanity – but of which science knows nothing.

Consequently, there can be no harmonious future for the civilization without a blending of these two critical streams.

34

Religion apart from science tends to become obscurantist, dogmatic and bigoted; science apart from some general view of human nature in its total context becomes meaningless and destructive. Unless science is practised on the basis of a conception of human nature that does justice to our highest aspirations, the prospect for the future is bleak indeed.

(Greene 1989)

I find it impossible to see a success of any scientifically-guided environmental management without a concurrent personal commitment motivated by moral obligations of responsibility and by willingness to share and to sacrifice. Science is no substitute for morality, and to believe that the ethics of limits and sharing has no place in dealing with our environmental dilemmas would be to forfeit any hopes for real successes in solving them.

WHAT ARE WE FACING?

Everything is tottering, and we have the impression of a particularly intense and acute movement of historical forces.
Nicholas Berdyaev, *The Meaning of History* (1936)

Confident diagnoses of the state of our environment remain elusive. Even where we have a fairly solid quantitative base, divergent interpretations can stretch the data in the direction of desired policy-making conclusions. These weaknesses are compounded in forecasts. We can simulate accurately many complex physical processes, and we can turn this ability into outstanding technical designs. But even our best simulations of long-term interactions among environmental, economic, technical and social developments have been simplistic and misleading. Biases have thus an enormous role in interpreting the past record and forecasting the future.

Perhaps, not surprisingly, intellectual reaction to degradative environmental changes has been found disproportionately in two antipodal camps. For catastrophists every severe pollution episode and every regional famine, every study of a chronic ecosystemic decay and every new appraisal of a declining resource base are merely predictable stones in a mosaic of inevitable doom. They see nothing but decay, decline and despair, and the most combative adherents of this depressive faith see no hope at all: they are immune to reality. Ehrlich (1968) concluded a generation ago that 'the battle to feed all humanity is over . . . nothing can prevent a substantial increase in the world death rate'.

Since Ehrlich's best-selling prognosis life expectancy at birth in the world's two largest poor nations, in China and India, rose from, respectively, about 58 to 69 years and from 42 to 57 years (World Bank 1991). Gaining 5–7 years of life expectancy per decade seems to me a very impressive

35

advance – but the great doomsayer has no apologies for his false predictions: he had merely supplanted *The Population Bomb* of the late 1960s with *The Population Explosion* of the early 1990s (Ehrlich and Ehrlich 1990). Other notable representatives of the catastrophic genre range from a merely sombre *Global 2000 Report to the President* (Barney 1980) to gloomy proceedings of the Cassandra Conference (Ehrlich and Holdren 1988) and to end-of-the-world scenarios of nuclear winter (Harvell 1986). A regular annual dose of catastrophism comes out in the *State of the World* compiled by the Worldwatch Institute.

Its president has kept on predicting the world at the brink of global famine since the early 1970s. In 1989 he concluded that 'the evidence leads me to believe that we have only a matter of years to get some of these environmental trends turned around' (Brown 1989). If he would be correct, there would be no hope for success: complex, inertial systems take much longer than just years to transform, especially when the requisite changes will have to be so all-encompassing, ranging from technical innovations with long lead times to major socio-economic rearrangements.

Antipodal to catastrophists are cornucopians, the techno-optimists revelling in large population increases as the source of endless human inventiveness who consider environmental decay as merely a temporary aberration. The most incorrigible exponents of this suasive, but naïve, idea went as far as claiming that with continuation of present trends the world of the year 2000 will be (emphases are theirs) *less crowded*, *less polluted* and *more stable ecologically* (Simon and Kahn 1984). To Simon both the numbers of people and mass of resources are basically immaterial: 'It is not resources per capita but rather free political and economic institutions that most strongly influence national wealth and that must logically be the focus of efforts to augment it' (Simon 1990).

A close reading reveals a wealth of irrefutable facts in the writings of both of these extreme groups – but these patches of sensibility are surrounded by a flood of intellectual bias precluding any recognition of a complex, largely unpredictable, and repeatedly counter-intuitive reality. The task is not to find a middle ground: the dispute has become too ideological, and the extreme positions are too unforgiving to offer a meaningful compromise. The practical challenge is twofold. First, to identify and to separate the fundamental, long-term risks to the integrity of the biosphere from less important, readily manageable concerns. The second task is to separate effective solutions to such problems from unrealistic paeans to the power of human inventiveness.

During the 1980s a good deal of research effort was started along both of these lines, and its conclusions are summarized in volumes by Repetto (1985), Malone and Roederer (1985), Clark and Munn (1986), Smil (1987), McLaren and Skinner (1987), DeFries and Malone (1989), Adams (1990), Silver (1990), Turner *et al.* (1990), Young (1990) and Mathews (1991).

Reliable determination of environmental impacts and understanding of their relative gravity should be much clearer with the progress of two major international research programmes launched during the 1980s: International Geosphere-Biosphere Programme (IGBP) initiated in 1983, and The Human Response to Global Change Programme (HRGCP) started in 1988.

And the late 1980s and the early 1990s also saw a flood of publications prescribing practical solutions to every conceivable environmental problem. These works range from consensus documents of international policy-making (IUCN/UNEP/WWF 1991) to illustrated popular reviews (Mungall and McLaren 1990; Sarre and Smith 1991), and from economic treatises (Pearce *et al*. 1991) to do-it-yourself guides for the greening of the Earth (Getis 1991). By far the most famous document in the first category is the Bruntland Report (World Commission on Environment and Development 1987) whose publication accelerated the process of the appropriation of global environmental change by politicians and government bureaucracies.

With all of this information on the record what is the justification of yet another look at environmental troubles and solutions? I have three simple answers. First, I will try to separate the essential realities of environmental needs and impacts from confusing myths. Such a start is the only sensible basis for informed decision-making. We have to deal with realities, and when our particular knowledge is inadequate we have to accept the necessities of uncertain probabilistic appraisals. Second, I will try to show that the still considerable gaps in our knowledge of the environment should not delay remedial actions. There is plenty of accumulated experience showing us how to act in effective ways. Essentials of such a strategy can be found in numerous writings long predating any interest in environmental affairs.

Third, I will try to demonstrate that while the potential for slowing down, stabilizing, and eventually reversing every degradative environmental trend is promising, the actual execution of such shifts will be generally very difficult. I believe that the overwhelming public approval for environmental improvements is based on misleading presentations of necessary changes. We are not facing marginal adjustments manageable by simple technical fixes, some temporary tax and spending increases, and a few changes of personal habits. We will need profound socio-economic transformation which will demand not only new ways of doing things but also not a few genuine sacrifices.

2

EXISTENTIAL NECESSITIES

HUMAN NEED FOR NATURAL RESOURCES

By the words, necessary of life, I mean whatever, of all that man obtains by his own exertions, has been from the first, or from long use has become, so important to human life that few, if any, whether from savageness, or poverty, or philosophy, ever attempt to do without it.

Henry David Thoreau, *Walden* (1854)

In 1845 Thoreau tested the extent of these necessities by building himself a house of ten by fifteen feet – 'the boards were carefully feather-edged and lapped, so that it was perfectly impervious to rain' – for a total cost of less than US $29, planting two and a half acres of potatoes and vegetables, and earning US $13.34 by surveying, carpentry, and various day-labour. His expenses, washing and mending aside, came to US $62 in eight months.

A lady once offered me a mat . . . I declined it, preferring to wipe my feet on the sod before my door. It is best to avoid the beginning of evil.

By such genuinely austere standards

the necessaries of life for man . . . may, accurately enough, be distributed under the several heads of Food, Shelter, Clothing and Fuel; for not till we have secured these are we prepared to entertain the true problems of life with freedom and a prospect of success.

Thoreau's definition is, of course, hopelessly unrealistic against the backdrop of modern western civilization whose driving force has been the incessant accumulation of goods and growing demand for services. But its very simplicity highlights the extent of our frivolous preoccupations by exposing the tremendous disparity between even very liberally interpreted necessities and the currently accepted norms and expected improvements of rich societies. At the same time, Thoreau's link between his minima and 'life with freedom and a prospect of success' reminds us that a very large part of mankind lives in circumstances in which even the true necessities are only precariously available or largely absent.

Both of these realities are fairly widely appreciated in basic qualitative terms, and much has been written about the potential for conservation and efficiency improvements in the rich countries, as well as about the existential misery throughout the poor world. But I have found repeatedly how shallow are the perceptions of these realities once one asks for quantitative details. Careful quantifications reveal not only astonishing misery and strained prospects but also greater possibilities for improvement than commonly realized. I will illustrate this by focusing on food, fuel and material needs, the foundations of decent human existence. Our dependence on these existential necessities has been also the principal cause of environmental degradation, and these impacts have been obviously magnified by a rapidly growing global population. Consequently, I will review briefly this enormous population rise before quantifying principal energy and material dependencies.

POPULATION, NUTRITIONAL NEEDS AND FOOD PRODUCTION

I think I can fairly make two postulata.
First, That food is necessary to the existence of man.
Second, That the passion between the sexes is necessary
and will remain nearly in its present state.
Thomas Robert Malthus, *An Essay on the Principle of Population* (1798)

Reconstructions of global populations indicate the total of less than 300 million at the beginning of the second millennium, rising to nearly 500 million by the year 1600 and surpassing 900 million at the very end of the eighteenth century (Carr-Saunders 1936). This progression translates into an average annual exponential growth rate of a mere 0.09 per cent for the first 600 years, and 0.3 per cent for the subsequent 200 years. By 1900 the total reached about 1.6 billion, at an average annual rate of nearly 0.6 per cent, and then it took only about 60 years of growth averaging 1.1 per cent a year to double to 3.2 billion (Figure 2.1). By 1990 the total stood at 5.3 billion and the forecast for the year 2000 is for about 6.3 billion, a total giving an average annual growth rate of almost 1.4 per cent during the twentieth century (UN 1991).

While it is easy to describe what has happened, it is difficult to appraise the role of individual factors. For most of human existence high fertilities were almost balanced by infant and childhood mortalities, resulting either in virtually stationary or only very slowly expanding populations. During the twentieth century fertilities throughout the poor world remained high, while better nutrition, clean water, urban sanitation, inoculation, antibiotics, and (more recently) large-scale food aid, brought impressive declines of mortality. This combination was the most important reason for nearly quadrupling

39

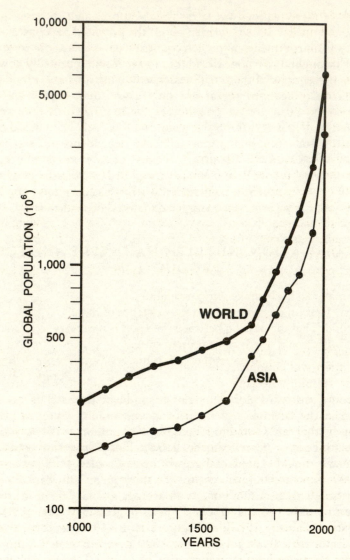

Figure 2.1 Exponential rise of global, and Asian, populations during the past two milleniums.
Source: Based on estimates in Carr-Saunders (1936), Demeny (1990), data in UN (1950–91), and forecasts in UN (1989).

the global population in a century. Even if the typical standard of living would have remained at the 1900 level and even if the population increases would have been rather uniformly distributed, this increase would have caused environmental changes of an unprecedented scale and intensity.

Uneven distribution of huge population increases has resulted in many extraordinary concentrations of environmental impacts. These changes were greatly intensified by advancing industrialization converting mostly rural, decentralized, parochial, low-energy subsistence societies into largely urban, centralized, interactive, high-energy cultures. This enormous socio-economic transformation is yet to be completed in many areas of the poor world. In continental terms, Latin America is most advanced on its path (with almost 80 per cent of the population in cities in 1990), Africa the least, with about 35 per cent of total population urban in 1990 (World Resources Institute 1992).

Population densities brought by urbanization are orders of magnitude above those of pre-industrial societies, and they have led to unprecedented densities of energy flows and raw material consumption. This intensive metabolism generates huge volumes of concentrated wastes, and poses unprecedented problems in removing and controlling a growing number of pollutants. Supplying cities with adequate food calls for large-scale food production dependent on high-energy subsidies, and causing major environmental changes ranging from growing water scarcity to the loss of bio-diversity.

Looking ahead remains surprisingly uncertain: long-range population forecasts during the twentieth century have repeatedly underestimated the coming global totals (Demeny 1989). But we may be more fortunate in forecasting the change during the next two generations when the global population will continue growing, but when its rate of expansion will be almost certainly slowing down. Average annual growth rates kept on increasing until the latter half of the 1960s when they peaked at about 2.1 per cent. Then, for the first time after millenniums of accelerating growth, the average global rate started to fall, to just short of 2 per cent during the early 1970s, to about 1.8 per cent during the latter half of the decade, and to 1.64 per cent during the 1980s. This truly revolutionary shift arose from a combination of below-replacement fertilities in most rich nations and, more importantly, from rapid fertility declines in many poor populous countries.

Most notably, China's fertilities were cut by some 60 per cent in a single generation (Banister 1988), and they are now approaching the replacement level. Rates are still considerably higher in other populous nations, but their declines have also been impressive (Figure 2.2). Transition to low fertilities is now well advanced, and the demographic consensus expects a further strengthening of this trend. This is a welcome shift, but one which will bring little near-term relief: the UN medium forecast puts the most likely time for reaching world-wide replacement fertility no earlier than in the third decade of the next century (UN 1991).

Even if this timing would not be delayed, the Earth would have an equilibrium population of around ten billion people, roughly double the 1990 total. Moreover, in the immediate future this trend will be overshadowed by the rising absolute additions: in the early 1970s the world was adding around

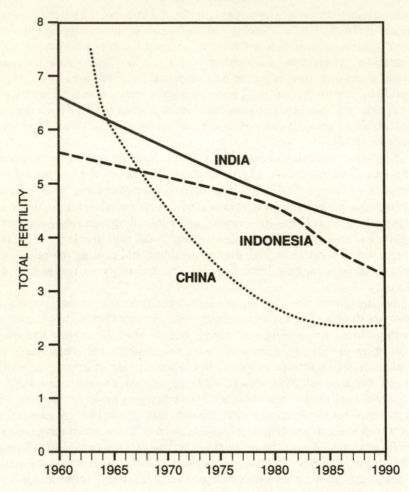

Figure 2.2 Declining fertilities in Asia's three most populous countries. The high Chinese rate of the mid-1960s was a compensation for very low (around three) fertilities during the famine years of 1959–61. Normal fertilities before the famine were around six.
Sources: Data from Smil (1992) and the World Resources Institute (1992).

75 million people a year, by 1990 the total rose to 87 million and it will average about 90 million during the 1990s, with more than nine-tenths in poor countries.

Food requirements and nutritional adaptations

Food consumption differences among rich and poor countries are both quantitative and qualitative. On the average, people in rich countries have an access to 25–50 per cent more food than they need for healthy and vigorous lives, but these rich diets are hardly desirable models. Average annual per capita meat consumption in most European countries and in North America surpasses normal adult body weights, most protein comes from animal foods, and fats and sugars are eaten in prodigious quantities (each year up to 30 kg of butter, lard and oil and as much as 60 kg of sugar per capita). In contrast, average food supply in most poor nations is just above the basic need, and typical diets are dominated by grain with small additions of vegetable oils or butter, relatively high intakes of vegetables, some fruit, very little sugar and only occasional meat, poultry, eggs or fish (in East Asia also hardly any milk).

Food energy can be derived from three different kinds of macronutrients. Carbohydrates are the most abundant products of photosynthesis and we can digest them only when they come as starches (polysaccharides) and simple sugars (monosaccharides fructose and glucose, and disaccharide sucrose). The poor world's carbohydrates come overwhelmingly from starches in cereals and from leguminous grains. Rich nations consume a large share of carbohydrates as refined sugar. Pure carbohydrates contain 17 MJ (megajoules) of food energy per kg, and both whole grain and milled cereals supply about 15 MJ/kg.

Proteins contain about one-third more food energy than carbohydrates (23 MJ/kg) but their primary role in human nutrition is as the source of amino acids essential to build and to repair body tissues (FAO/WHO/UNU Expert Consultation 1985). We cannot synthesize proteins without consuming first eleven essential amino acids in plant and animal foods. These acids must be also present in correct ratios, and all animal foods and mushrooms supply such perfect proteins. All plant foods are deficient in one or more essential amino acids, but complete proteins can be supplied by combining cereal grains, always short in lysine, with legumes which are short in methionine but have plenty of lysine.

Lipids (fats) have also important non-energy roles: their essential fatty acids must be consumed preformed in either plant or animal fats, and they also carry fat-soluble vitamins (A, D, E and K). Fats, with 39 MJ/kg, more than twice as energy dense as carbohydrates, produce readily the feeling of satiety and make bulk carbohydrates more palatable. Fat supply remains low in subsistence farming societies (rarely above 15 per cent of the total food intake), but it has risen to between 30–40 per cent of the total in industrialized nations. Compared to the three macronutrients, vitamins and minerals are needed in minuscule amounts but their deficiencies prevent normal metabolism and growth. Balanced diets usually supply adequate amounts of both of these micronutrients but deficiencies are common even in rich nations, especially among children, young adults and elderly in low-income groups.

Diets throughout the poor world are overwhelmingly vegetarian, dominated by cereals, legumes and tubers. Several nearly universal dietary changes accompany improvements in the average standard of living as declining intakes of whole grain cereals, tubers and legumes are replaced by higher consumption of animal foods, refined sugar and fresh fruits. Affluence brings lower carbohydrate and higher lipid and protein shares, but some vitamin and mineral deficiencies may persist. As for the total energy intakes, all industrialized nations have average per capita food availabilities (mostly between 12.5–15 MJ/day) much above the actual average needs of less than 11 MJ/day. Storage, retail and kitchen waste makes up the difference. In contrast, even in the better-off poor countries the average food availability is only marginally above the need.

Given the generally unequal, both in spatial and in socio-economic terms, food distribution, even such countries have large shares of their populations affected by various degrees of malnutrition most commonly noticeable in children. For example, China's average food availability is now roughly 10 per cent above the estimated need, but during the late 1980s nearly at least 100 million people had access to less than 9.2 MJ of food energy a day, the minimum level necessary to meet FAO's primary nutritional objective of healthy and vigorous life (Smil 1993). And in 1990 about one-fifth of China's children were malnourished, and two-fifths suffered from moderate to severe stunting (Grant 1990).

All estimates of national or global nutritional deficiencies remain highly uncertain because determination of human energy needs is complex and because people can adapt to lower food intakes. Human food needs consist of the basal metabolic rate (BMR), energy needs for growth and replacement of body tissues, and mark-ups for activities. BMR is the minimum amount of energy needed at rest for maintaining the critical body functions and it varies with sex, body-size and age, and it has substantial departures from statistically expected means (FAO/WHO/UNU Expert Consultation 1985). Daily BMRs of adult men fall mostly between 6–9 MJ, for adult women between 5–7 MJ. Minimum survival requirements call for at least 25 per cent more energy for metabolizing food and for personal hygiene even for housebound people.

Additional energy is necessary for childhood and adolescent growth and during pregnancy and lactation. Growth demands account for as much as one-third of the total food intake during the first few months of life, and after an early teen rise to 3 to 4 per cent they become insignificant. Energy cost of pregnancy equals between 15–20 per cent of BMR, much less than the cost of subsequent lactation (30–50 per cent of BMR for most western women). But these results do not seem to apply to the poor world's pregnancies. Studies from several poor countries show that pregnant women can give birth to healthy babies while consuming commonly 20–40 per cent less food energy than expected on the basis of the western experience (Prentice 1984). Similarly large discrepancies were established for breastfeeding women. Higher metabolic efficiencies induced by limited food availability are the best explanation.

Energy costs of activities can be expressed conveniently as multiples of BMR. They range from 1.8–4.0 BMR for light work and to 6–8 BMR for heavy exertions. Much of fieldwork in traditional farming societies still calls for heavy labour, while a great majority of tasks in industrialized societies needs merely light exertions. Mental work increases the BMR by less than 10 per cent. Estimates of total food energy requirements of a particular population cannot be done without some simplifying assumptions. FAO's figures of daily per capita rates at the beginning of the 1990s ranged from just over 9 MJ for India to about 11 MJ for the USA. Per capita means for poor countries, with their relatively young populations and low body-weights, are mostly between 9.6–10 MJ/day, for the rich nations they cluster around 10.8 MJ.

Clear evidence of successful adaptations to reduced food availability means that considerable metabolic variability exists not only at individual but also at group level: there is no single pre-set minimum of food energy supply applicable to all populations (Borrini and Margen 1985). Workers with low-energy intakes are often as productive as those with high food consumption, long-term adaptation to lower food availability (achieved by slowing of growth and reduction in adult body mass) can maintain good health (albeit in smaller but energetically more efficient bodies), and there is hardly any correlation between food energy intake and time spent actually working (Edmundson and Sukhatme 1990).

Better understanding of food energy needs and human adaptabilities to limited food intakes has resulted in a declining trend of global estimates of undernourished people. In the 1950s FAO put the number of people with inadequate food intakes at over half of the world-wide population, while the World Bank estimated that during the late 1980s about one-third of all people in industrializing countries had inadequate food energy intakes (Reutlinger and Pellekaan 1986). Even the latter share is most likely still a substantial overestimate.

Drawing the lines for undernourishment largely on the basis of western nutritional research, and setting single reference values for population energy needs ignores the existence of considerable human variation and of effective physical and social adaptation to lower energy intakes. Of course, these adaptations are not without costs: they limit the body's fat reserves and weaken the chances of coping successfully with additional external stress. Undernutrition exists, but its frequency and impact are much less important than those of malnutrition. We should be concerned much more with the quality of the diet than with the quantity of food intake.

Moreover, as Edmundson and Sukhatme (1990) conclude, 'it is unwise and perhaps even immoral to examine the complex problems of human health only in terms of human energy needs . . . More food is not the key to development'. Still, more food will be needed for growing populations, and it can be produced by expanding the area of cultivated land, by intensifying the cropping of existing farmland, or by a mixture of the two basic strategies.

Expanding crop harvests

The unprecedented twentieth century's rise in the global food output has come from a combination of extended cultivation and intensified cropping. Only an approximate apportioning of the two contributions is possible. The global area of cropland was about 1.5 Gha (billion hectares) in 1990, while the uncertain 1900 total was as little as 900 Mha (million hectares) or as much as 1.1 Gha, corresponding, respectively, to expansion rates of 36 and 67 per cent. As the annual harvest of key crops expanded at least fivefold, cropping intensification was responsible for more than four-fifths of this large gain. Poor but land-rich societies always prefer expansion, and the largest 20th-century gains (more than doubling the cultivated area) have been in Africa, South Asia and Latin America. In contrast, China's cultivated land grew by less than one-third, and Europe's remained virtually constant (Richards 1990).

Possibilities for further extension of cultivated land remain enormous: the global total of potentially cultivable land is approximately twice as large as the total area cultivated in 1990. But a highly unequal distribution means that Africa could expand its farmland almost fivefold while the reserves are small in Central America and in Europe, and marginal in many populous Asian countries (Dudal 1987). When a further expansion of cultivated land is either physically impossible, too inconvenient or too expensive, then only intensification can support larger populations. Its traditional forms range from simple ditch irrigation to such elaborate solutions as extensive terracing and rotations of staple cereal crops with legumes grown for food or as green manure.

Its modern ways take the form of higher energy subsidies which come directly as fuels and electricity powering field machinery, irrigation and crop-drying, and indirectly as chemical fertilizers, pesticides, machines, implements, tools and improved seeds. Higher subsidies applied to more responsive crops – above all short-stalked varieties of wheat introduced during the early 1960s (Hanson *et al.* 1982) and hybrid corn (Sprague 1977) – resulted in substantially higher yields. Average global harvests of the two leading staple grains, wheat and rice, have been growing steadily since 1950 and comparisons of these means with best national averages indicate a large potential for further improvements (Figure 2.3). Maxima harvested at experimental stations or during trials attempting to set record yields range between 12–14 t/ha (tonnes per hectare) for wheat and 9–13 t/ha for rice, approaching the highest theoretical yields of, respectively, 18 and 14 t/ha (Gilland 1985).

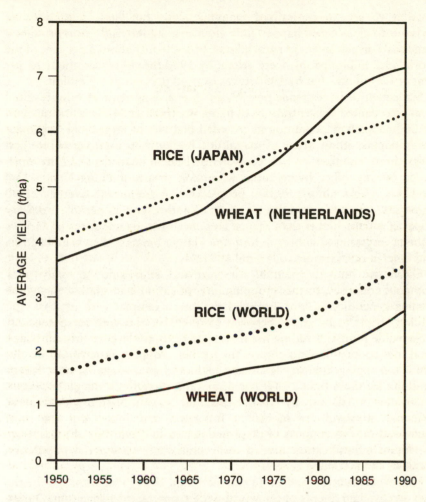

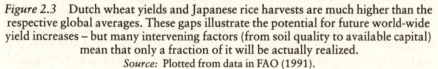

Figure 2.3 Dutch wheat yields and Japanese rice harvests are much higher than the respective global averages. These gaps illustrate the potential for future world-wide yield increases – but many intervening factors (from soil quality to available capital) mean that only a fraction of it will be actually realized.
Source: Plotted from data in FAO (1991).

Progression of maximum population densities brought by rising energy subsidies has been impressive. Shifting agriculture – still practised by at least 100 million Asian, African and Latin American families – can feed typically just 20–40 people/km^2, traditional cropping supports mostly 100–800 people and modern farming can feed up to 2000 people/km^2 of cultivated land. This intensification also dispenses with large numbers of the rural labour force and it accelerates the process of urbanization which, in turn, stimulates more concentrated ways of cropping. Fossil energies made it possible to set up centralized manufacturing, but mass migrations of peasants to cities could happen only after farming intensification brought a large-scale displacement of rural labour. In 1800 only about 5 per cent of the world's 1.2 billion people were urban; by 1900 the share was about 15 per cent of 1.7 billion – but by 1990 it was over 40 per cent of 5.3 billion.

Settlement densities made possible by intensive agriculture have resulted in unprecedented concentrations of humanity (Hall 1984). Densities around 10,000 people/km^2, common in crowded quarters of large cities, represent anthropomass in an order of magnitude higher than total vertebrate and invertebrate zoomass of rich ecosystems. Peak human densities are unrivalled by any other living form. Mongkok, heart of Hong Kong's Old Kowloon, packs almost 90,000 people/km^2. Assuming an average of 40 kg/person, this rate translates to 3.6 kg of anthropomass per m^2, a density about 200 times higher than that of large herbivorous ungulates in Africa's richest ecosystems and three to four times larger biomass than in all bacteria and fungi in rich farming soils (Smil 1991a).

Given the limited potential for farmland expansion in many poor populous countries, further cropping intensification is inevitable during the coming generation. China provides the best example: with at least 300 million people to be added between 1990–2010, and with the potentially reclaimable farmland adding less than 10 per cent to the currently cultivated total, the country has no choice but further cropping intensification. But China's cropping intensity is already high, and its average farming energy subsidies (at more than 25 GJ (gigajoules)/ha three times the global mean) are rivalled in Asia only by Japan, Korea and Taiwan. Nation-wide totals of potential farmland are higher in Indonesia and India, but the most intensively cultivated parts of these two countries (Java, Punjab) anticipate the future intensification needed to accommodate substantial population increases of the coming generation.

Higher fertilization and more extensive and better irrigation will be the two decisive ingredients of the world-wide cropping intensification. During the next few generations synthesis of nitrogenous fertilizers will not be limited by the availability of the best feedstock (natural gas) and deposits of phosphates and potash will not be in short supply either (Sheldon 1987). In contrast, many regions and countries are already approaching the limits of their sustainable water withdrawals, and irrigation is always responsible for most of these needs.

Photosynthesis is an inherently water-intensive process. Depending on climate and crop species (C_3 varieties, the majority of staple crops including wheat and rice, are considerably less water-efficient than such C_4 cultivars of which corn and millet are the only major grain plants), 1 kg of grain requires between 600 and 1600 kg of water (Doorenbos et al. 1979). In contrast, even such water-intensive industrial processes as wood-pulping or chemical syntheses require typically just between 50–250 kg of water per kg of finished product, and daily residential needs are mostly between 150–400 kg per capita (Kammerer 1982). Not surprisingly, water withdrawals for irrigation average almost 90 per cent of total water use in Asia and Africa, three-fifths in South America and half in North and Central America (World Resources Institute 1992).

Even if water availability would be of no concern, continuing intensification of cropping means that future food supplies will be even more dependent on extraction and conversion of fossil fuels, on generation of electricity, and on the supply of numerous minerals: without these critical inputs there would be no production of inorganic fertilizers and pesticides, no mass manufacturing of field machinery and irrigation pumps and pipes, and no ways of reliable transportation and efficient processing.

ENERGY RESOURCES AND CONVERSIONS

It is important to realize that in physics today, we have no knowledge of what energy *is*.
Richard P. Feynman, *The Feynman Lectures on Physics* (1963)

But there can be no doubt about the practical consequences of high rates of energy conversions: nothing sets the modern industrial civilization so much apart from its traditional agricultural predecessors as the huge flows of energies supporting the well-being of an average person. A pre-industrial European consumed annually mostly between 20–40 GJ of wood and charcoal for cooking, heating and manufacturing, and the maxima in North America of the nineteenth century went as high as 70–100 GJ/capita. With average fuel conversion efficiency no higher than 15 per cent these rates equalled three to six, and exceptionally 10–15 GJ of useful energy. In contrast, in 1990 the world's rich nations consumed about 150 GJ of primary energy per capita. With average conversion efficiency of at least 40 per cent this is about 60 GJ of useful energy, an order of magnitude above the typical rates of the biomass era.

Biomass combustion was much lower in pre-industrial Asia, Africa and Latin America, ranging mostly between 5–20 GJ/capita, and, with conversion efficiency of just around 10 per cent, equalling no more than 1 to 2 GJ of useful energy. This disparity intensified during the twentieth century, as direct consumption of fossil fuels is still below one GJ/capita in the poorest African countries, and it averages less than 25 GJ/capita for all industrializing nations. Consequently, global distribution of fossil fuel consumption remains extremely skewed.

In 1990 the rich nations with less than a quarter of the world's population consumed about three-quarters of all primary energy (the USA alone, with less than 5 per cent of global population, used 25 per cent of the total), while the poorest quarter of mankind – made up of some 15 sub-Saharan African

49

countries, Nepal, Bangladesh, Indo-China and most of rural India – consumed less than 3 per cent of the world-wide flow of commercial energies.

Every terrestrial civilization must be always fundamentally a solar one, depending on the sun's radiation to maintain a habitable troposphere and to drive photosynthesis. In pre-industrial societies almost instantaneous solar energy conversions supplied not only all the food, but also all fuel and mechanical power (animate labour, water and wind) employed in farming, construction, transport and manufactures. Even with the introduction of more efficient tools and machines, all traditional solar societies had to work within rather limited energy budgets. As a result, the growth of their productive capacities, as well as of their populations, had been very slow.

Extensive extraction and conversion of solar energy accumulated and transmuted over millions of years in fossil fuels brought an enormous increase of useful power capacities resulting in a civilization profoundly different from its solar predecessor. Its three key characteristics are a plentiful food supply, unprecedented material affluence, and high personal mobility. The first achievement is made possible only because of large and growing inputs of fossil energies into agriculture. The second one rests on large amounts of energy invested in production of such basic commodities as metals and building materials as well as in provision of highly energy-intensive services. High mobility of people using private or public means has been made possible by the introduction of progressively more powerful, lighter and more efficient prime movers ranging from steam-engines to gas turbines.

Fossil fuels and electricity have also virtually eliminated human labour as the source of kinetic energy. Mechanization of agricultural and industrial tasks has changed the human role in productive tasks to that of a controller and manager of inanimate energy flows. This has freed an increasing amount of time for leisure, much of it filled with pastimes requiring additional inputs of fossil energies and materials.

Fossil-fuelled civilization

Quantitative advance of fossil-fuelled civilization can be charted rather accurately almost from its very beginnings. All important coal-mining nations have fairly good output statistics going back to at least 1850, and nineteenth century hydrocarbon production was recorded even more carefully (US Bureau of the Census 1975; Mitchell 1975). In contrast, biomass energy uses during the past century can be only roughly estimated, but the best evidence points to the 1890s as the great energetic watershed: by 1900 the gross energy content of fossil fuels equalled that of all biomass energies. Given the higher conversion efficiency of fossil fuels, this means that the twentieth century advances were energized from its beginning largely by fossil fuels. During its last decade these fuels provided an order of magnitude more energy than biomass.

A semi-logarithmic plot of global fossil fuel production shows a hundred-fold rise from about 2.6 EJ (exajoules) (82.5 GW (gigawatts)) in 1850 to just over 300 EJ (nearly 10 TW (terawatts)) in 1990 broken into four distinct periods (Figure 2.4). After three generations of a nearly perfect exponential rise (4.3 per cent/year until the beginning of the First World War) came a spell of slow increases (until 1932) followed by a resumption of vigorous growth (4.5 per cent/year) lasting until 1974. Since that time growth rate

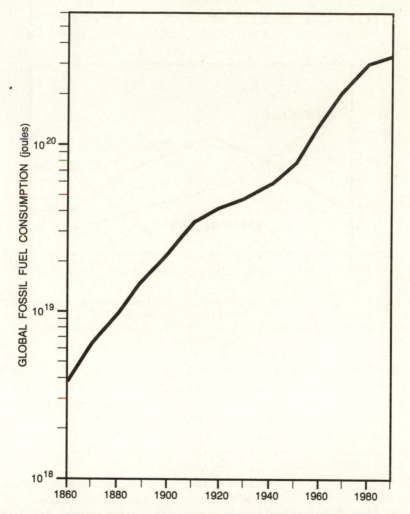

Figure 2.4 Exponential rise of global fossil fuel consumption has been the critical factor in developing and advancing modern industrial civilization.
Source: Based on Smil (1991a).

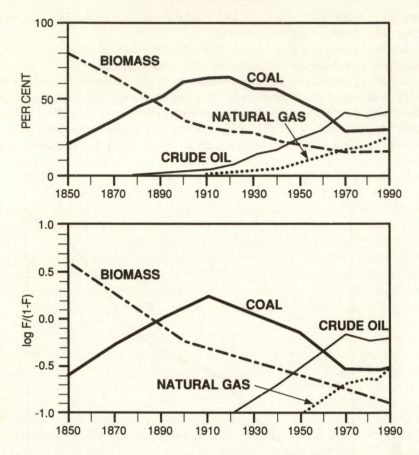

Figure 2.5 Orderly substitutions of leading primary energy resources show long lead times required for a new source to gain a substantial share of the market.
Source: Based on Marchetti and Nakicenovic (1979) with updates for the 1980s from UN (1980–91).

averaged just 0.7 per cent/year (UN 1991). Both the output and consumption of fossil fuels have been always highly concentrated in a small number of countries. The British lead ended in the early 1890s and the USA has been dominant ever since. Russia is the second largest producer, and China surpassed Saudi Arabia's falling output by 1981 to become the world's third largest producer of fossil fuels. In 1990 these three largest producers extracted half of the world's fossil fuel total (BP 1991).

The huge production increase has been accompanied by a world-wide transition from solid fuels to hydrocarbons, which surpassed coals' contribution by 1960, and has been providing about two-thirds of all fossil energies since 1970. Substitutions are slow, with about a century needed to capture half of the market (Figure 2.5) In spite of many factors influencing the production the penetration rates remain constant over long periods of time (Marchetti and Nakicenovic 1979). Transitions on national level are much less orderly. Future transition will certainly remain slow and any suggestion of an abrupt end to the fossil fuel era are naïve.

Another important qualitative change has been the rising share of fossil fuels used indirectly as electricity. Reasons for this trend are compelling. Only electricity offers the following combination: instant, effortless consumer access; ability to fill every conversion niche and be turned into motion, heat, light and chemical potential with easy control and unmatched precision and speed; silent, clean (at the point of final conversion) and highly reliable delivery; and capacity for easy accommodation of growing or changing uses. Moreover, electricity can be converted to heat with nearly perfect efficiency, it can deliver temperatures higher than in combustion of any fossil fuel, and its use does not need any inventory! All of these attributes make electricity the preferred choice for countless final uses.

Expansion of the world-wide electricity generation from less than 10 TWh (terawatthours) in 1900 to nearly 12 PWh (petawatthours) in 1990, more than a thousandfold rise, has been one of the most important technical triumphs of the century leading to rapid increases of turbogenerator and plant ratings, introduction of higher transmission voltages, and emergence of international systems (Smil 1991). Thermal electricity accounted for about three-fifths of the total in the 1920s, three-quarters in 1970 and 64 per cent in 1990 (UN 1991). With an average conversion efficiency of about 33 per cent, the 1990 world-wide thermal generation consumed roughly a quarter of all extracted fossil fuels. In 1990 USA produced almost exactly 30 per cent of the 7.6 PWh of global thermal generation and the largest producers (USA, Russia, Japan, China, Germany and UK) accounted for nearly 70 per cent of the world total.

Growing demand for fossil fuels and electricity has led to increased unit sizes, be it for individual techniques or complete production systems. In turn, these advances often stimulated an even higher demand, and led to a gradual concentration of extraction, processing and conversion industries. Such concentrations necessitated the development of extensive transportation and distribution networks delivering fuels and electricity over progressively longer distances and creating strong intra- and international dependencies. In 1990 energy exports included one-tenth of all hard coal, nearly two-fifths of crude oil and one-seventh of natural gas output, altogether about 80 EJ (2.5 TW). Oil trade is a truly global undertaking and several principal export streams (Persian Gulf–Japan, Middle East–Europe) deliver 10^{11} W over distances up to 12,000 km (Figure 2.6).

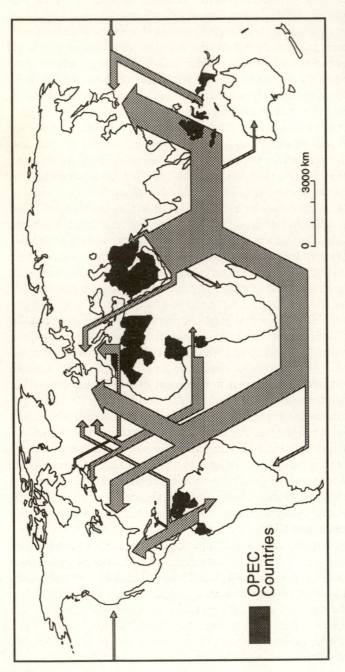

OPEC
Countries

0 _____ 3000 km

Figure 2.6 Major crude oil export streams show the dominance of Middle Eastern supplies and the global nature of the oil trade. Thickness of arrows corresponds to shares of global shipments during the late 1980s.
Source: Based on OPEC (1991).

Non-fossil contributions

Traditional biomass energy sources continue to make essential contributions in most poor populous countries, and two non-fossil conversions, hydro and nuclear electricity generation, are prominent in many industrialized and in some industrializing nations. Accurate appraisals of the global dependence on biomass fuels are impossible as the poor families harvest themselves the bulk of the supply. National estimates indicate dependence in excess of 80 per cent for much of sub-Saharan Africa, about 50 per cent in Indonesia, over 30 per cent in India, and more than 20 per cent in China and Brazil.

Field surveys in wood-burning societies give equivalents of between 0.5–2.5 m^3 (5–15 GJ) of air-dry wood a year per capita. Where peasants depend largely on crop residues and dried dung the rates are much lower, just between 2–8 GJ a year per capita. Supply shortages have been affecting increasing areas on all three poor continents. China's rural energy surveys put the average daily effective heat energy requirement at 16.2–18.7 MJ per family, but the actual availability was a mere 14.5 MJ a day, an average shortfall of over 20 per cent (Smil 1988). Seasonal and regional shortages are commonly more than twice as high. Other acutely affected areas include Africa's Sahelian zone and Namibia, Swaziland, Lesotho and parts of Botswana, the Nepali hills, large parts of the Indian subcontinent, and much of Central America (FAO 1980).

During the late 1980s the annual biomass consumption of poor nations was about 25 EJ, with China, India, Brazil and Indonesia being the largest consumers. Fuelwood consumed in the rich countries for household heating and by forestry and pulp and paper industries added about 5 EJ, for a grand total of around 30 EJ/year, an equivalent of almost exactly 10 per cent of the world-wide consumption of primary commercial energies.

Large-scale hydroelectric generation started before the Second World War with state-supported projects in the United States (Tennessee Valley Authority) and in the USSR. By 1990 there were more than 120 stations with capacities in excess of one GW in operation in more than 30 countries (the largest station, Itaipu on the Paraná between Brazil and Paraguay, was designed for the ultimate power of 12.6 GW). The United States has both the largest number of such plants and the largest installed capacity (about 70 GW). Canada, Russia, and China rank next.

In 1990 hydro generation accounted for just over 20 per cent of world-wide production, and in many African and South American countries water-power generates more than 80 per cent of all electricity. In terms of oil fuel equivalents (converting to a common denominator by calculating the amount of fuel burned in oil-fired plants in order to generate the same amount of electricity) hydrogeneration of electricity contributed about 540 Mtoe (million tons of oil equivalent) in 1990, almost 20 per cent more than nuclear generation, but less than 7 per cent of the total consumption of primary energy.

The world's first nuclear station, British Calder Hall, came on line in 1957. By October 1973, when OPEC embarked on a series of sharp oil price rises, there were more than 400 reactors in operation, under construction or in various planning stages in 20 countries. Yet instead of taking advantage of higher oil price, nuclear industry went into a retreat. Escalation of construction cost and public distrust were the two major reasons for the stagnation and decline (Rossin 1990). The perception of intolerable risks was strengthened first by the 1979 Three Mile Island accident and then by the 1986 Chernobyl disaster (US Department of Energy 1987). The lack of progress in solving the matter of long-term disposal of radioactive wastes has further lowered the chances of public acceptance. Still, in 1990 this nuclear generation provided almost 20 per cent of the world's electricity, with the highest contributions in France (more than 70 per cent), Belgium, Taiwan, South Korea and Sweden.

Other ways of electricity generation – by wind, the Earth's heat, ocean tides and waves, and direct (photovoltaic) or indirect (central tower) solar energy conversion – will continue making some local differences during the 1990s, but their shares of global primary energy supply remains negligible, and fundamental physical considerations of low power densities will limit their future expansion (Smil 1991a).

PRINCIPAL MATERIAL NEEDS

At the present day, and in this country, as I find by my own experience, a few implements, a knife, an axe, a spade, a wheelbarrow, etc., and for the studious, lamplight, stationery, and access to a few books, rank the necessaries, and can all be obtained at a trifling cost.

Henry David Thoreau, *Walden* (1854)

We know of no other life than the carbon-based biota of the Earth created by photosynthesis. Powered by solar radiation, this conversion forms complex organic compounds from CO_2 and H_2O, and from much smaller inputs of the three macronutrients (nitrogen, phosphorus and potassium) and a dozen micronutrients (ranging from calcium and sulphur to cobalt and zinc). Grand biogeochemical cycles of water, carbon (involving tropospheric CO_2, organic matter and sedimentary carbonates), nitrogen (between atmospheric N_2, NO_x and N_2O, and nitrates and ammonia in soils and proteins in biomass) and sulphur (mainly as SO_x, H_2S and sulphates), and mobilization of the other nutrients by microbial metabolism, soil weathering and erosion, have assured global abundance of these indispensable inputs during life's evolution.

Traditional farming supplemented these natural flows by irrigation and by recycling of organic wastes, and pre-industrial societies added a number of new requirements by extracting growing volumes of building materials

56

(wood, clay, stone, sand) and metals (copper, tin, zinc, iron, silver, gold). Still, as Thoreau maintained, material necessaries of life in excess of food, shelter and fuel could amount to very little and could be had at a small cost. Industrial advances changed that. Mass manufactures and steam- and electricity-driven mechanization greatly expanded the need for traditional raw materials, and added a large number of minerals, ranging from silica to vanadium, to civilization's basic needs. But compared to similarly rapid expansions of food and energy production, there are two very important differences in assessing our material needs and their environmental impacts.

First are the known magnitudes of all but a few non-fuel mineral resources. Huge increases of material inputs supporting the world-wide industrialization brought recurrent concerns about the adequacy of mineral resources needed to meet future supplies (President's Materials Policy Commission 1952; Meadows *et al.* 1972). Fears about depleting some metallic ores have been especially prominent, but misplaced: virtually all minerals have experienced long-term declines in real prices during the last two generations. In contrast to the finite amount of farmland and to the limited deposits of easily extractable liquid and gaseous fossil fuels, resources of nearly all non-fuel minerals are so enormous that even growing demand would not exhaust reserves of any critical element before the middle of the twenty-first century.

This abundance allows for gradual introduction of new techniques which exploit lower grade ores and alternative compounds which could bring some 30 key elements into practically infinite supply (Goeller and Zucker 1984). Of course, mineral resources are finite in a global sense, but from a civilizational perspective 'resources are not; they become,' and 'the process of becoming is one that is as much ideational as it is material' (De Gregori 1987). The second principal difference is a ready substitutability of many mineral resources. While life has its immutable chemical requirements, and while the eventual replacement of fossil fuels by solar conversions cannot be done without a profound social transformation, many materials can be replaced easily by cheaper, more durable or environmentally more acceptable alternatives.

An outstanding example of these possibilities are the ebbing uses of copper, the first important metal of ancient civilizations, one of the mainstays of pre-industrial manufactures, and an essential ingredient of industrialization. The metal can be replaced by aluminum in electrical equipment, car radiators and refrigerator tubing, by titanium in heat exchangers, by plastics in pipes and plumbing fixtures and, most importantly, by glass fibres in cables (Weisser *et al.* 1987; Desurvire 1992). Between 1975 and 1992 the transmission capacity of optical fibres has increased tenfold every four years (Figure 2.7). By the mid-1990s novel types of optical amplifiers, based on a fibre doped with erbium and powered by laser diode chips, will make it possible to carry simultaneously 500,000 phone calls per cable!

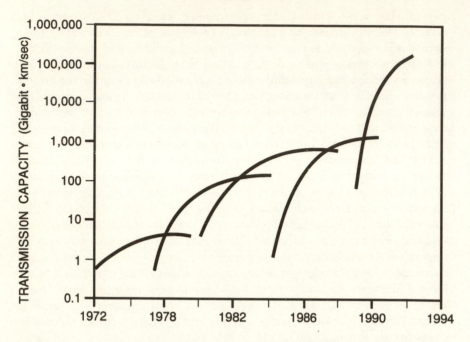

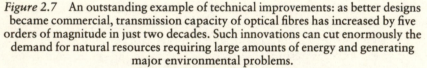

Figure 2.7 An outstanding example of technical improvements: as better designs became commercial, transmission capacity of optical fibres has increased by five orders of magnitude in just two decades. Such innovations can cut enormously the demand for natural resources requiring large amounts of energy and generating major environmental problems.
Source: The graph is derived from Desurvire (1992).

Iron, used largely as steels, remains the principal metal of industrial civilization – its annual global output in 1990 was nearly 13 times as large as the combined total of five other leading metals, aluminium, copper, zinc, lead and tin – but its world-wide production has been stagnating for nearly a generation. This reflects not only substitution (by plastics, ceramics and composite materials) and conservation (lighter finished products and more durable steels) but also of ownership saturation in most rich nations where products embodying the largest amount of steel – cars and large domestic appliances – have hardly any new markets left (Larson *et al.* 1986).

Among the non-metallic minerals phosphorus occupies a special place: its relative scarcity and indispensability make its long-term availability a matter of world-wide concern. With nitrogen abundant in the atmosphere (making the synthesis of ammonia largely a matter of sufficient energy for splitting gaseous N_2 and providing cheap hydrogen), and with potassium salts plentiful in the lithosphere, phosphorus is the least abundant macronutrient. Neither complex sugars nor proteins can be made without it because their syntheses are powered by energy released by the phosphate bond moving reversibly between two phosphate compounds (Westheimer 1987).

Unlike carbon, nitrogen and sulphur, phosphorus is not volatile and so its biospheric flows have no significant atmospheric link from ocean to land. On a civilizational time-scale of 10^3 years there is no P cycle, merely a one-way flow with interruptions moving the element from soils and rocks into the ocean (Smil 1990). Continuation of present growth in consumption of fertilizer phosphates would exhaust all of the element's reserves in about 50 years, and all suspected resources in less than 250 years (Sheldon 1987). This will not happen because both the extent of farmland and optimum fertilizer applications are limited. Consequently, in the long run the fertilizer needs will follow a linear trend and the present reserves may last over 200 years, and identified economic resources may be sufficient for almost two millenniums (Fox and Yost 1980).

Moreover, the traditional belief that the major part of fertilizer phosphorus is quickly tied up by the soil in forms unavailable to crops has been a costly misconception: most agro-ecosystems actually have P utilization efficiencies between 70–90 per cent (Karlovsky 1981; Smil 1990). Applying phosphorus to crops is not thus largely an expensive exercise in transferring the element in soluble phosphates to insoluble soil compounds.

While the rich nations have already moved well into the era of declining relative material needs (reflected in at least stabilized and often declining absolute consumption), the poor world's growing populations aspiring for higher standard of living will keep up the global demand for materials for a number of generations. But during the next generation this reality should not encounter either any acute resource constraints or generate any unmanageable environmental consequences.

3

ENVIRONMENTAL CHANGE

HUMAN TRANSFORMATION OF THE BIOSPHERE

Should the planet become uninhabitable before mankind has reached
maturity . . . the world's effort fully to centre upon itself could only be
attempted again elsewhere at some other point in the heavens.
                    Teilhard de Chardin, *Man's Place in Nature* (1966)

The often repeated arguments about billions of galaxies full of stars suitable
to hold in orbit planets much like our own make an eminent probabilistic
sense – but they do not mean the inevitability of billion-fold replications of
the Earth's life. I am impressed less by the possibility of countless
civilizations (including many vastly more sapient worlds than ours) sprin-
kled through the cosmic void than by the idea of the Earth being the only
harbour of life in the universe. In any case, until we can get in touch with
an extraterrestrial society, we should behave as if the existence of life
confined to a single planet is the only reality. Then, if we are to remain and
to prosper on this planet, we must reconcile our needs with biospheric limits.
Our actions will continue changing the environment, but we must, not
before too long, arrive at levels of interference which are compatible with
long-term preservation of critical biospheric functions.

We are still moving away from this goal, but we have at least started to
realize the enormity of environmental transformation which is imperilling
the survival of modern civilization. These changes can be ordered into three
broad categories: declining availability of critical natural resources and
services; changing composition of the atmosphere; and the loss of biodiversity. The first change will make it impossible for the majority of the
global population to achieve even incipient economic prosperity during the
next few generations – and increasingly difficult to support the quality of life
now expected by inhabitants of rich nations. Although these declines apply
to a wide variety of resources, diminishing availabilities of productive
farmland, fresh water and mature natural forests must be by far the greatest
concerns.

The second transformation alters the radiation balance of the atmosphere, as well as the composition of the stratosphere, and the presence of trace compounds in the troposphere. The third degradation reduces variety of the Earth's life. This change undermines the effective functioning of indispensable biospheric services (as interdependence of species is among the hallmarks of biotic complexity), and it destroys the irreplaceable wealth of genetic information which could have potentially enormous value in improving human well-being. Before proceeding with systematic critical reviews of these changes I will first look at their spatial extent and frequency. Such a classification exercise is essential in order to appreciate the impacts of these transformations and the public and policy-making concerns they generate.

DIMENSIONS OF ENVIRONMENTAL CHANGE

Establish order before confusion sets in.

Lao Zi, *The Way of Life*

The simplest way to classify environmental change is by the area affected. This common denominator makes it possible to trace the evolution of degradative effects and to compare consequences of disparate processes. The extremes are from local to global, the intermediate steps from small to large regions to semi-continental and continental effects. Approximate quantifications are best done in terms of the orders of magnitude. Local effects are limited to areas of 10^0 km^2, small regional impacts to 10^1–10^2 km^2. Large regional changes affect mostly between 10^4–10^5 km^2, the semi-continental ones 10^6 km^2 (Figure 3.1). Examples for most individual categories abound.

Air and water pollution and land dereliction which accompany the early stages, as well as the subsequent intensification, of industrialization energized by coal are classical symptoms of local environmental impacts. Most importantly, they include high ground-level concentrations of particulate matter and sulphur dioxide, acid mine drainage, and the disposal of coal-mining, sorting and washing wastes. England of the 1880s or North China of the 1980s would be perfect illustrations. Other examples range from destructive slopeland erosion brought about by poor agronomic practices to high levels of untreated urban waste in rivers.

Growth of industrial conurbations and of fossil fuel industries extended a number of environmental impacts to regional level. In poor countries improper intensification of farming and overexploitation of grasslands and forests had the same effect. By the 1960s notable manifestations of these impacts could be seen on every continent. In America they included photochemical smog covering the Los Angeles Basin and strip-mining destruction in Appalachia. In Asia the affected regions ranged from Tokyo Bay overloaded with industrial waste water to massive deforestation in Indian and Nepali Himalayas. Among the worst examples in Europe was the

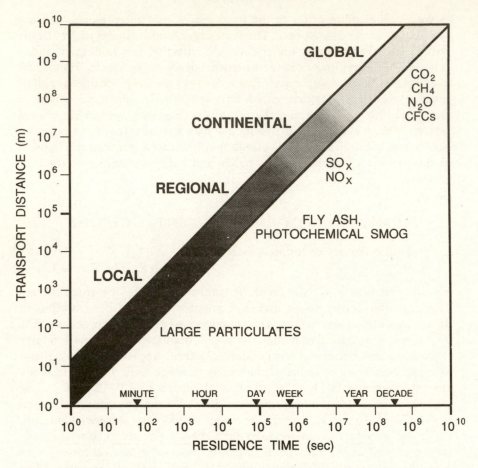

Figure 3.1 Spatial and temporal scales of environmental change. Local to global effects are illustrated by various forms of gaseous and solid emissions to the atmosphere.

land destruction and exceptionally high SO_2 emissions in coal-mining regions of former East Germany and Czechoslovakia. In Africa the damage extended from over-irrigated and salinizing fields in the Nile delta to the deforested and eroding slopelands of Rwanda and Uganda.

By the late 1960s came a clear identification of the first semi-continental environmental effect as measurements showed that a large part of Europe – from Scandinavia to northern Italy, and from the Netherlands to southern Poland – is subject to increasingly acid deposition (Figure 3.2). Soon afterwards the same condition was confirmed for eastern North America (Figure 3.3). The late 1960s also saw a dramatic extension of catastrophic pollution to the open ocean. In contrast to safe and unobtrusive pipeline

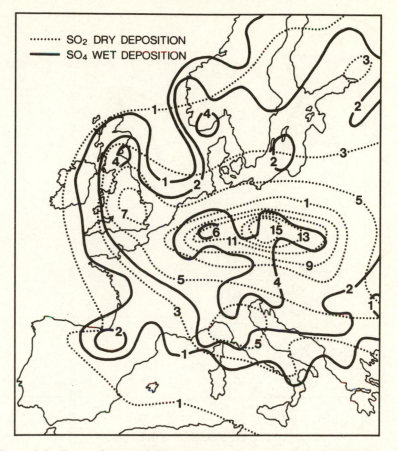

Figure 3.2   Dry and wet acid deposition over Europe in the mid-1970s, before most of the continent's nations agreed to reduce their emissions. Deposition rates are expressed, respectively, as SO_2 and SO_4 in g/m^2.
Source: Based on Ottar (1976).

transportation, the growth of tankers – from less than 20,000 t in the late 1940s to over 300,000 t by the early 1970s – increased the risk of crude oil spills. The *Torrey Canyon* disaster in 1967 was the first large (more than 100,000 t) oil spill, followed by a number of supertanker accidents, and the blow-up of a well in the Bay of Campeche released about 1.4 Mt (millions of tonnes) in 1979–80 (Teal and Howarth 1984).

Increased attention to renewable energy sources brought a greater realization of often considerable environmental problems associated with large hydro stations. Those located on lower courses of rivers, where turbines operate under relatively low heads but with huge volumes of water, create large reservoirs. The largest ones – Akosombo on the Volta in Ghana (8730 km²) and Kuybyshev on the Volga (6500 km²) – are almost as large

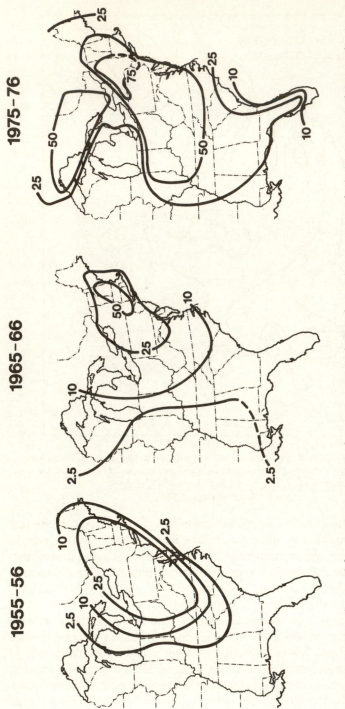

1955–56   **1965–66**   **1975–76**

Figure 3.3   Spreading acidification in eastern North America. Annual rates show concentrations of H⁺ in microequivalence per litre. *Source:* Based on Likens and Butler (1981).

as small countries (Lebanon, Cyprus or Puerto Rico). Inundation of alluvial land is the most contentious environmental change when planning new reservoirs. Silting is the most worrying degradation in the completed ones, but in warm climates there are additional concerns about the spread of infectious diseases and for the uncontrolled growth of water weeds (International Commission on Large Dams 1978; Ramberg *et al.* 1987).

During the 1970s it also became clear that excessive soil erosion is more than just a limited regional problem. With virtually every intensively cultivated hilly or mountainous region experiencing unsustainably high soil losses these problems grew to become large regional, or semi-continental threats. And the 1980s were marked by a sudden rise of worries about the two indisputably global environmental risks: possibility of a relatively rapid climatic change caused by growing atmospheric concentrations of anthropogenic greenhouse gases, and a major loss of the stratospheric ozone brought about by the emissions of long-lived chlorofluorocarbons.

But this useful classification is of little help in assessing the relative importance of environmental impacts. To begin with, the effects are parts of continua, and their extent depends on classifications which are either basically arbitrary (is the baseline for acid deposition pH 5.6 or 5.0?; is the tolerable concentration of SO_2 50 or 20 μg/m^3?) or have to be adjusted in complex ways to account for different ecosystemic realities (sustainable erosion rates will differ with soil type, air pollution damage to plants will vary with dominant species).

And the global change is not necessarily more threatening than a combination of regional degradations, especially if the latter ones are very widely distributed. This reality requires a specific labelling of the two classes of global problems. For a common class of environmental changes affecting nearly every major region of the ecumene (inhabited part of the earth) – excessive soil erosion, depletion and pollution of aquifers, deforestation, or irrigation-induced salinization – whose cumulative impact already has ecosystemic as well as socio-economic consequences, but whose progression in one region has little or no direct near-term impact on a situation elsewhere, I will use the adjective world-wide. For those interferences whose effects may have a planet-wide impact on civilization and biota I will reserve the label global. The Earth's changing radiation balance and a relatively rapid decline of stratospheric ozone are obviously two foremost examples in this category.

Human interference in the three grand biogeochemical cycles of carbon, nitrogen and sulphur generates effects ranging from local to global. Burning of tropical rain forests destroys more than the local biodiversity, and changes potentially much more than the local climate: with many endemic species the decline in biodiversity may represent an irreplaceable global loss, and the CO_2 released by burned trees may change the global climate. Interference in the nitrogen cycle has a clear global component owing to the increase of atmospheric levels of N_2O, a greenhouse gas generated by denitrification of nitrogenous fertilizers which can also affect stratospheric O_3 levels.

But most near-term nitrogen-related concerns are largely local or regional. Besides denitrification, fertilization results in higher volatilization and leaching. Leaching, together with erosion, can cause contamination of groundwater and eutrophication of surface waters. The other two major changes of local and regional importance are from the concentrated generation and discharge of organic N compounds by urban populations and confined livestock, and from emissions of nitrogen oxides during stationary and automotive fossil fuel combustion.

Some environmental impacts may arise rather rapidly: excessive water withdrawals from large aquifers (North China Plain, Ogallala Great Plains) or the emergence of intolerable photochemical smog (Mexico City, Taipei) came about in less than a generation. And some world-wide degradations may prove to be global within a few generations: deforestation and erosion are the two most obvious possibilities.

DECLINING AVAILABILITY OF NATURAL RESOURCES

The wheels and springs of man are all set to the hypothesis of the permanence of nature. We are not built like a ship to be tossed, but like a house to stand.

Ralph Waldo Emerson, *Nature* (1836)

Environmental changes have been increasingly a source of local or regional resource shortages, and have been weakening the productive capacity of affected areas. But until recently these transformations were not matters of national interest in any large populous country: while some of their regions always suffered, such countries could behave as if the critical resources would be always available. They could bring more of the less accessible land into production, could turn to more remote forests, tap more distant water resources or deeper aquifers. Simply, they stood and grew on the apparent permanence of natural goods and services.

During the past two generations the combination of large population increases and rising economic expectations did away with this illusion. The addition of at least three billion people in industrializing countries before the year 2025 guarantees that the rate of anthropogenic environmental stresses will keep on increasing even if the average standard of living would be no higher than today's poor world average. But this will not be the case: average per capita advances may be much slower than desired or expected, but further gradual increases in food intake, water use, and energy and raw material consumption will translate into greater environmental impacts, and hence into further loss and degradation of natural resources.

Modern economic and policy-making focus has been always on possible

66

resource shortages, but we do not need the vast majority of plants and heterotrophs as commodities. Rather, the most fundamental aspect of environmental degradation is the irreplaceable loss of existential services provided by biota. Plants maintain the unique composition of the Earth's atmosphere by photosynthetic reduction of CO_2, and act as a huge filter of airborne particulate matter and a major sink for gaseous air pollutants. They also help to regulate the planetary water-cycle through evapotranspiration and storage (both in living tissues and in litter), and they prevent or moderate floods by reducing excessive water runoffs.

Bacteria perform the two key conversions in the nitrogen cycle by transforming ammonia into nitrates (nitrification) and returning nitrate nitrogen to the atmosphere as N_2O and N_2 (denitrification); they also synthesize ammonia from atmospheric dinitrogen (fixation). Bacteria, together with fungi and soil invertebrates, also form irreplaceable links in the global carbon cycle: they generate and maintain productive soils by decomposing voluminous organic wastes. Soil heterotrophs also recycle plant nutrients, aerate soils and store water, and degrade many chemical pollutants (pesticides, alkaline and acidic compounds, hydrocarbons) both in contaminated soils and waters. As predators or parasites living organisms are also responsible for controlling most pests, and insects and some vertebrates are irreplaceable pollinators. And, of course, the biota store and perpetuate the enormous amount of genetic information needed for effective adaptations.

Balanced evaluations of degradations diminishing the resource base, and imperilling the maintenance of biospheric services, are difficult. There is a great deal of obvious destruction and deterioration, but reliable quantitative appraisals of national, and even more so of world-wide, losses are largely missing. Closer looks at some changes also indicate that the standard, usually highly alarmist, perceptions may be exaggerating the real rate, as well as the eventual extent, of environmental degradation.

Farmland changes and soil erosion

Losses and degradation of arable land are world-wide phenomena, caused by growing populations, industrialization, and environmental changes brought largely by improper agronomic practices – but they are most acute in many poor populous countries (Blaikie and Brookfield 1987; Dudal 1987). Perhaps the most stunning fact is that China's annual farmland loss averaged almost exactly 1 Mha (million hectares) between 1957 and 1980, and although the yearly rate has been declining recently, a conservative forecast for the 1990s indicates a cumulative loss of at least another five Mha, an equivalent of all farmland in Hungary, or a third of Indonesia's total. In the absence of any major opportunities for farmland expansion (even very costly reclamation could enlarge China's cultivated area by less

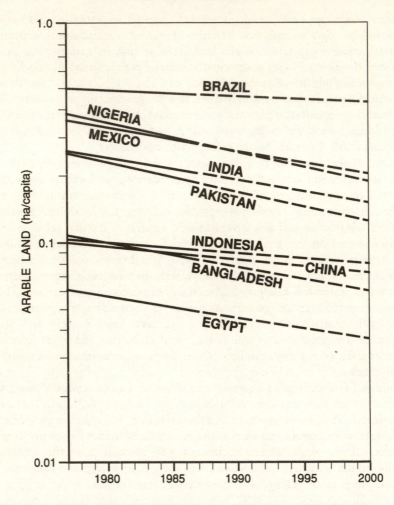

Figure 3.4 With Brazil as the single exception, all poor populous countries have a relatively small amount of farmland which is steadily declining. Rates below 0.1 ha/capita require high intensity of cultivation to maintain largely vegetarian diets.
Source: Based on farmland and population statistics in FAO (1991) and population forecasts in UN (1991).

than 10 per cent of the current total) such losses can be made up only by further intensification of cropping.

And the Chinese are far from alone in this unenviable position. FAO's (1981) detailed appraisal of long-term agricultural prospects in 90 poor countries (excluding China) found that 51 of them had abundant or moderately abundant reserves of arable land – but their population was just one-third of the assessed total. In contrast, 18 countries with extreme land scarcity, already cultivating an average of 96 per cent of the potentially arable land, supported half of all the population (Figure 3.4). With reclamation opportunities virtually gone, and with growing space requirements for larger populations and higher industrial production (often claiming the best alluvial soils) these countries face continuous declines in per capita availability of arable land.

Housing of an additional one billion people during the 1990s and expansion of the requisite industrial production and transportation will add up to large farmland losses, especially in India, China, Nigeria, Pakistan, Bangladesh, Brazil and Indonesia, the nations where half of the projected increase in the world population is to occur during the next generation. Growing cities will also require more space for solid waste disposal and for water treatment. And agriculture itself is a substantial consumer of arable land: irrigation needs about 0.1 ha of land for access, canals, pumping stations and pipes for each ha of effectively watered land.

The only way to increase harvests from diminishing areas of farmland is to intensify cropping. In populous poor countries with small average holdings this does not mean large-scale mechanization, but further extension of irrigation and, above all, much higher applications of synthetic fertilizers, above all of nitrogen. Traditional inputs of the nutrient – recycling of organic wastes, planting of legumes, and cultivation of nitrogen-fixing *Azolla* in paddy fields – can supply as much as 200 kg N/ha for multicropping (Smil 1991b). This input will produce annually a maximum of 200–220 kg of protein per ha. With overwhelmingly vegetarian diets this is enough to sustain 12–13 people, prorating to a minima of 700–800 m^2 (0.07–0.08 ha) of arable land per capita.

Regional and national averages – reflecting the limiting effects of inadequate water, poorer soils and pests, as well as the necessity to cultivate non-food crops – will have to be progressively lower. For example, China's double-cropping rice region as a whole averaged less than seven people per ha, and the nation-wide mean for traditional farming was about 5.5 people/ha, or 0.18 ha/capita. But in 1990 China's mean was at best 0.1 ha/capita, and at least 400 million people lived in provinces having less than 0.08 ha/capita. Clearly, even the most assiduous traditional multicropping unlimited by water supply or major pest damage and natural disasters could not produce enough staple grain for the basic subsistence with so little land. Not surprisingly, China has become the world's largest producer and user of synthetic nitrogen, with annual applications averaging almost 200 kg/ha, roughly four times the US mean.

High population densities – generally in excess of four to five people per ha in temperate climates with rainfed single-cropping, and above ten people in warm, irrigated multicropped regions – are thus supportable only by high applications of nitrogenous fertilizers supplemented by smaller amount of phosphorus and potassium. World-wide synthesis of nitrogenous compounds rose from 3.6 Mt in 1950 to just over 30 Mt in 1970 and to about

90 Mt in 1990, and in the early 1990s about half of the nutrient was applied in poor populous countries (FAO 1990). Further large increases are inevitable.

A realistic short-term need can be estimated by assuming (conservatively) the following decadal changes: a 5 per cent decline in the typical yield response (higher applications bring steadily diminishing returns); a 10 per cent increase in total per capita protein intake in the poor world; and a 10 per cent decline of arable land per capita in the land-scarce countries containing half of the world's poor population. These shifts would increase the annual demand for synthetic nitrogen by at least 15 Mt in the year 2000, and 30 Mt a decade later, bringing the respective totals to around 115 and 145 Mt N. The latter amount would be roughly a 70 per cent increase in a single generation, a change intensifying the concerns about the environmental effects of synthetic nitrogen.

To begin with, synthesis of ammonia even in the most efficient plants requires at least 30–35 GJ/t (gigajoules per tonne) of nitrogen, and the subsequent formulation of urea raises the cost to 50–80 GJ/t of nitrogen (Mudahar and Hignett 1987). Higher fertilization will thus also increase the need for fossil fuel and electricity. Leaching of nitrates into aquifers and streams will almost certainly increase, a link demonstrated by the rising nitrogen loading of the Mississippi. The river's lower-course nitrate level remained fairly constant between 1905 (the year of the earliest measurements) and the early 1970s, but it increased more than fourfold between 1970 and 1985. The river's watershed receives about 40 per cent of all fertilizer applied in the US and the rising applications have been responsible for roughly three-quarters of the increased nitrogen burden (Turner and Rabalais 1991). Fortunately, the total loading (about 1.5 mg/l) is still well below the maximum acceptable level.

The situation is clearly different in Europe and Asia. Royal Society Study Group Staff (1984) found nitrate levels exceeding allowable limits in many English localities. Nitrate concentrations in the Dutch drinking groundwater started to rise sharply in the early 1970s, and a decade later they were just above the European Community standard of quality (25 mg NO_3/l, 2.5 times above the US standard), and by the mid-1980s they were surpassing 40 mg NO_3/l (Ministerie van Volkshuisvesting, Ruimtelijke Ordening en Milieubeheer et al. 1985). Regular monitoring would undoubtedly reveal high nitrate levels in both surface and groundwater of heavily fertilized rice regions of Asia. And much as in the Dutch case, there will be numerous new local and regional 'take-offs' of nitrate concentrations throughout the populated areas of the continent during the coming generation.

Earlier concerns about nitrogen-induced eutrophication – enrichment of surface waters with nutrient runoff from fields – eased after extensive research had confirmed that the surfeit of phosphorus, rather than the presence of nitrates, is responsible for excessive algal growth (OECD 1982).

A different kind of worry now focuses on volatilization of ammonia from fertilized farmlands and also from livestock. According to Nihlgard's (1985) hypothesis, these atmospheric inputs may be largely responsible for excessive fertilization of some forests by nitrogenous compounds.

Cropping intensification has not been paid for only with added energy cost of synthetic fertilizers and with nitrates in waters. Synthetic nitrogenous compounds cannot maintain soil's high organic matter content and its appropriate structure promoting crop fertility, moisture retention and rich soil fauna. This has been a world-wide problem, but it is especially acute in many poor countries where the less common cultivation of green manures is combined with the declining recycling of crop residues and with less frequent applications of fermented wastes. The organic matter content of many farming soils is now just around 1 per cent, compared to the desirable range of 2 to 3 per cent.

The low content of organic carbon reduces the numbers of earthworms, the principal soil tillers and irreplaceable agents of soil humification (Hartenstein 1986). They turn over soil, bury surface litter, increase soil aeration, convert organic nutrients, promote the growth of nitrogen-fixing bacteria and destroy some phytopathogenic fungi – but their optimum growth requires heavy recycling of organic wastes. Hardening of soils and higher runoffs are the other frequent consequences of this deterioration. But the leading world-wide concern associated with intensified cropping is the extent and intensity of soil erosion.

Erosion has several obvious degradative effects: it reduces the soil's organic matter and fine clay content, diminishes its water-retention capacity, removes crop nutrients, and limits plant rooting depth (Harlin and Berardi 1987; Lal 1990). This degradation leads to a progressive decline of average crop yields, but there is very little reliable evidence to quantify these losses (Schertz *et al*. 1985). American experiments with corn and soy bean grown on plots with artificially removed topsoils demonstrated variable yield declines in trials without fertilizers – but no difference when adequate nitrogen was added. But these results come from a small number of short-term experiments and cannot be used to generalize on long-term impacts.

In the rich countries the agronomic advances – above all steady increases in fertilizer applications, introduction of new cultivars and better field equipment – have so far easily masked all but the most extreme (and hence localized) negative effects of soil erosion. In the future it may not be so easy to replicate this record, but we simply do not know. This admission is based on a little appreciated fact: even for the richest countries we have only rough estimates of total soil mass eroded (Higgitt 1991), and our quantifications of actual nutrient losses are no more than informed guesses. Among the large agricultural producers only the USA have instituted a periodic survey of soil erosion as a part of the National Resources Inventory (NRI).

The first sampling, in 1977, did not include wind erosion losses, but the

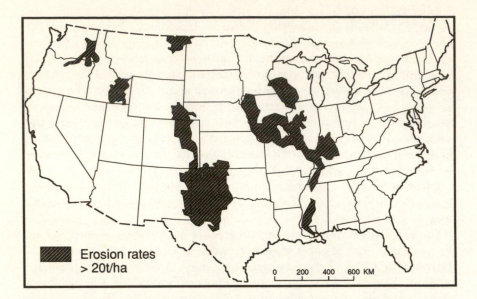

Figure 3.5   Areas of the highest soil erosion in the United States, the country with
the best periodic surveys of wind and water erosion, are concentrated on the dry
wheat-growing Southern Plains and in parts of the wet Corn Belt.
Source: Based on Lee (1990).

1982 and the 1987 NRI, based on nearly one million sample points, gave a
comprehensive account (Lee 1984 and 1990). In 1982 water erosion
averaged about 9.6 t/ha and wind erosion amounted to 7.1 t/ha for a total
of 16.7 t/ha and (on 170 Mha of cropland) an aggregate annual loss of over
2.8 Gt (billion tonnes). By 1987 the water erosion mean was down to 9.1
t/ha, with declines occurring in every region of the country, but wind erosion
losses rose slightly. Relative severity of these losses can be appreciated by
comparing the mean with maximum average annual losses compatible with
sustainable, productive, and economical cropping. The highest tolerance
rate is about 11 t/ha, and during the 1980s over 40 per cent of America's
cropland was eroding at rates exceeding sustainable maxima (Figure 3.5).

But it would be highly misleading to believe that a sensible indication of
fertility decline could be obtained by simply multiplying the eroded mass by
a typical nutrient content. In order to estimate nitrogen lost in eroded soil
we should know first the mass of the nutrient present in the field (it can vary
considerably within even a small area), average enrichment ratio (topsoil
carried away will have usually much higher N content than deeper layers),
the share of the element available to support crop growth, rates of natural
replenishment (in precipitation and acid deposition and from soil mineral-
ization), and the degree to which eroded soil merely displaces the nutrient
to another part of the field.

As already noted in the first chapter, only the first variable is fairly well known for fields with repeated soil testing: for the rest we cannot usually offer even good estimates. This is especially true about the degree of erosive nutrient translocation, a critical value in determining the net nutrient loss. In gently-sloping croplands most of the eroded soil does not end up in waterways, but simply on different patches of cultivated land. A careful assessment in an area of small relief without any major surface outlet (landscape type common in north-central USA) showed that very little or no eroded sediment may be leaving the cultivated area (Larson *et al.* 1983). In contrast, improperly farmed fields on steeper slopes may lose most of their eroded soil into streams.

These large local differences make it very difficult to estimate actual nutrient (or organic matter, or water-retention capacity) losses, but Crosson (1985) concluded that for the USA both the past and the prospective effects of soil erosion on crop production costs are smaller than most of the soil conservation experts have long believed. For example, in the cornbelt, America's most productive crop region, continuation of erosion losses prevailing during the late 1970s would lead to average yield losses of just about 4 per cent in 100 years, with the range from 2 per cent on very gentle slopes to 18 per cent on steeper fields (Pierce *et al.* 1984). But Croson (1985) also added that 'although small, the costs probably are higher than in the public interest they ought to be'. And so even for the USA the best available long-term assessment ends with an uncertain judgement!

In the case of grossly destructive losses observable in such erosion-prone regions as China's loess plateau or parts of South Africa there can be no doubt about high economic costs, but regional or national estimates of total soil loss in Asia, Africa and Latin America are no better than order of magnitude guesses. World-wide erosion costs, ecosystemic and economic, are most likely unacceptably high, but in most cases reliable quantification of gradual losses, needed in order to assess cost-benefit ratios of remedial measures, remains elusive.

Forests, grasslands, wetlands

Large-scale deforestation marked the rise of dynastic China, the growth of Mediterranean civilizations, early European wood-based industrialization, as well as North American and Russian expansion during the nineteenth century. But during the twentieth century most of the forest losses happened in Asia, Latin America and Africa, and since the 1970s they have been overwhelmingly concentrated in the largest remaining areas of tropical rain forest, in Amazonia and in South-east Asia. Although the peak losses during the 1980s almost certainly set new annual records, comparisons of best available estimates and surveys show that even these rates are not so different from historical experience.

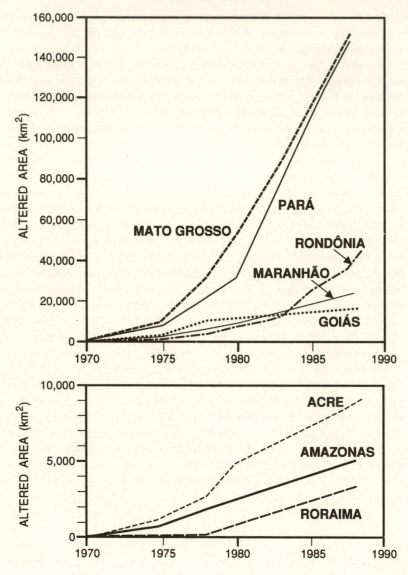

Figure 3.6 The best estimates and satellite surveys show that deforestation in Brazilian Amazonia increased rapidly during the late 1970s and the early 1980s in every state, but Amazonas, the core of the region, is still relatively untouched. The most recent surveys show a clear moderation of annual deforestation rates.
Source: Adapted from Fearnside (1990).

Just over 800 Mha were cleared world-wide between 1850 and 1980 at an average rate of roughly 6 Mha/year (Williams 1990a). Lanly's (1982) comprehensive survey of tropical deforestation set the annual rate of clearing at 7.5 Mha for all closed forests (whose canopies cover at least 20 per cent of the ground). Myers (1980) put the losses at 22 Mha, but his estimate was for forest conversions including fallow area within closed forests, and appropriate adjustment brings it very close to Lanly's total.

Inter-American comparison also reflects these similarities – as well as critical differences. In North America about 65 Mha of forests and woodlands were cleared between 1860 and 1980, the total virtually identical with the Latin American clearance (Williams 1990a). The two very similar averages of about half a Mha a year hide two very different timings. North America's most intensive deforestation phase was over by 1900 as the formerly large contiguous Eastern forests were reduced to a pattern of isolated patches. Since then these forests have been actually taking over abandoned low-quality farmland, while Latin America's deforestation continued during the twentieth century, and reached new peaks during the 1980s. These losses, especially in Brazilian Amazonia, attracted much western attention and condemnation.

Problems with the interpretation of remote-sensing data from satellites (the only viable method of relatively rapid assessment of large-scale deforestation) led to both overestimates and underestimates of annual losses. Fearnside's (1990) critical evaluation of satellite data shows that, within the area of legal Amazon (total of 4.975 million km^2), the annual deforestation rate peaked in 1987 at about 3.5 Mha, and that the total between 1970 and 1988 came to about 36 Mha, with four-fifths of the cleared forest in Mato Grosso and Para (Figure 3.6). According to Williams (1989) the same area of forest was cleared in the United States between 1780 and 1850. Given the large difference in population – US had just over 23 million people in 1850, Brazil over 150 million in 1990 – recent Brazilian deforestation rates have been actually much lower in relative terms than the American deforestation before the middle of the nineteenth century.

In historical perspective Brazil's large recent forest losses do not appear at all excessive, and the western moralizing about Brazilian destructiveness loses much of its superiority and contributes little to international understanding and co-operation. Stating this is not in any way approving the recent Brazilian practices. They differ from North American, as well as from earlier European deforestation, in two important ways. First is the setting: species-rich tropical rain forest growing typically on poor soils, an ecosystem where most nutrients reside in the biomass whose destruction leads to massive erosion, and whose regeneration is extremely slow – compared with temperate forest composed of a relatively small number of tree species growing on generally better soils covered by a nutrient-rich litter layer which reduces erosion and helps in a relatively fast re-establishment of new growth.

Second are the causes of deforestation and its long-term impacts. While most North American conversions resulted in the establishment of pro-ductive cropping, it is generally impossible to supplant the Amazonian forest with permanent fields, and cropping is limited only to a few years.

Consequently, most of the Amazonian forest destruction has not been caused by clearing for farming, but rather by conversion to large-scale grazing promoted by government subsidies (Mahar 1989; Binswanger 1991). Cattle pastures are yet another transitory plant formation whose exploitation stops with excessive erosion, invasion of inedible species or diffusion of palm trees (Fearnside 1990). In contrast, recent deforestation in Southeast Asia has been caused largely by commercial logging, conversion to tree plantations, and in Indonesia also by government-sponsored resettlement schemes.

Simple projections of high clearing rates of the mid-1980s would see all tropical rain forests gone well before the end of the next century. Similar projections done in 1860 would have showed no North American forests remaining before the end of the twentieth century. Tropical forests will not disappear before 2100 because in some countries (Guyanas, Zaire) clearing proceeds slowly and it is unlikely to escalate soon, and in other countries deforestation rates will eventually come down, and larger areas will be protected. The trend has already changed in Brazil, from the average of just over 2 Mha/year in 1985–9 to about 1.3 Mha in 1990 and just over 1 Mha in 1991 (Rohwer 1991). Even reduced losses could eliminate most forests outside protected areas within the next two to three generations. The environmental effects of this pantropical transformation include some indisputable changes as well as a few uncertain consequences (Mather 1990).

The two potentially most important global impacts are the extremely rapid loss of biodiversity and large-scale release of CO_2. A typical hectare of Amazonian forest contains more than 500 plant species and thousands of different heterotrophs and its burning and subsequent decomposition will release, assuming an average of 400 t of standing phytomass, about 660 t of CO_2. Destruction of forests with high shares of endemic species has been the most important reason for biodiversity loss during the past two decades, and the annual CO_2 emissions from forest burning during the 1980s were at least 700 Mt and possibly as much as 2 Gt (Houghton and Skole 1990).

Deforestation also brings much higher erosion rates and hence high nutrient losses. Differences between treed and burned or logged sites may easily surpass an order of magnitude, and the resulting higher siltation worsens the effect of downstream floods (Smil 1993). Deforestation also increases average water yield and river flow, but this sponge effect makes little difference in the case of major floods. The uncertainties include effects of large-scale deforestation on precipitation patterns as a result of decreased evapotranspiration and changed surface albedo.

There is little doubt about the importance of forest evapotranspiration, as well as the evaporation from open water bodies, for maintaining Amazonia's water cycle (Victoria et al. 1991) – but long-term effects of massive Amazonian deforestation remain uncertain. Nobre et al. (1991) concluded that the reduction in precipitation after replacing the Amazonian forest by grasslands would be larger than the projected decrease in evapotranspiration, and that a complete and rapid deforestation could be irreversible in the southern part of the basin. In contrast, Henderson-Sellers and Gornitz (1984) believe that there will be no significant effects on local rainfall.

There have been some sizable increases in national totals of temperate forests: a 70 per cent rise in Switzerland between 1863 and 1983, a 40 per cent expansion of French forests during the same period, and a nearly 50 per cent extension in Hungary between 1925 and 1980 (Mather 1990). But both the composition and the health of temperate (and also many boreal) forests have changed profoundly with massive logging of unprotected old growth and with creeping degradation caused by a combination of natural stresses (disease and insect outbreaks, drought, windstorms) and human impacts, above all by acid deposition. Effects on forest soils include loss of alkaline cations and mobilization of aluminum and heavy metals – but natural soil formation processes are frequently a more important cause of acidification than the deposition of sulphates and nitrates (Krug and Frink 1983).

Higher acidities, especially in parts of Central and Northern Europe, undoubtedly cause greater environmental stress, but even there it is incorrect to see a simple link between acid deposition and forest decline. Acid precipitation is just a part of the syndrome of *Waldsterben* caused by extreme weather (windstorms, drought) pests, and high levels of ozone and heavy metals (Hinrichsen 1986; Klein and Perkins 1987). Complexity of this degradation and its cumulative nature makes it virtually impossible to make a clear quantitative linkage between forest damage and distant emissions of acidifying substances or heavy metals (Bormann 1990). As with the creeping soil losses, this reality precludes the meaningful benefit/cost evaluations needed to justify investment in appropriate controls.

Grasslands have neither the species richness nor the large standing phytomass of forests, but their world-wide extent now surpasses the area of closed forests, and their high productivity accounts for one-quarter of all photosynthetic energy conversion (Smil 1991a). Grasslands, rather than forests, support the highest densities of mammalian zoomass on the planet, are still the basis of subsistence for millions of pastoralists in Africa and Asia, and protect many vulnerable soils from rapid erosion. Main human interferences undermining their productivity, and even their survival, are overgrazing, seasonal burning, and desertification.

Together with forests, wetlands are the most productive ecosystems: net primary productivity of many marshes easily matches that of tropical rain forest growth. Wetland detritus offers plentiful nutrition for bacteria, fungi, protozoa and invertebrates which provide the foundation for high-yielding aquatic trophic webs made up of numerous arthropod, fish, amphibian, avian and mammalian species. Wetland habitats are irreplaceable for a large number of commercial fish species as well as for waterfowl and for many migratory birds. Besides being rich reservoirs of life, wetlands' other benefits range from flood mitigation (by temporarily storing runoff waters) and coastal protection (by absorbing waste energy and reducing erosion) to aquifer recharge and sediment trapping, including considerable interception and processing of pollutants and toxic residues (Williams 1990b). They also act as a large carbon sink.

Once again, our inadequate knowledge makes it difficult to say what shares of wetland species and major ecosystems have already disappeared and what portions are threatened. Even the estimates of wetlands existing in the USA, the country with the most extensive surveys and assessments, varied nearly twofold, between 21 and 37 Gm^2 during the 1980s (Hofstetter 1983). Corresponding uncertainties must be much greater for largely uninventoried wetlands of the three poor continents.

Finally, I should mention the continuing destruction of coastal habitats colonized for at least three billion years by microbial mats. Deceptively unproductive, many barren coasts, evaporite flatlands and sandy intertidal zones shelter complex, multilayered communities of aerobic and anaerobic bacteria which have been instrumental in sequestering enormous masses of carbon by precipitating carbonates and iron in banded formations (Margulis and Guerrero 1989). Habitats of these essential biogeochemical processors are disappearing through coastal construction, reclamation and pollution.

Water resources

Very much like the topsoil loss, declining availability of water has become a world-wide phenomenon. L'vovich *et al.* (1990) estimate that the planetary river runoff declined by about 6 per cent (or nearly 2500 km^3) during the past 300 years as the total water withdrawal rose from just over 100 to about 1400 km^3 by 1950 and to more than 3600 km^3 a year by 1980. The largest decreases – brought by a combination of consumptive uses in crop irrigation, human and livestock consumption and industrial production – were in Asia (about 70 per cent of the total volume) and in North America.

But appreciation of this critical environmental change cannot be gained from aggregate statistics. Nation-wide rates of the intensity of water utilization indicate critical stresses only for a handful of densely populated countries in arid environments (World Resources Institute 1992; Shuval 1987). Countries of the Arabian peninsula can support their economic expansion only with expensive desalinated water because their demands are now greatly surpassing the maximum renewable supply of fresh water (Figure 3.7). But on regional scales declining water availability is now a problem in every large country, including the nations with enormous total water flows, and severe local water shortages make life difficult for tens of millions of poor people in Africa and Asia.

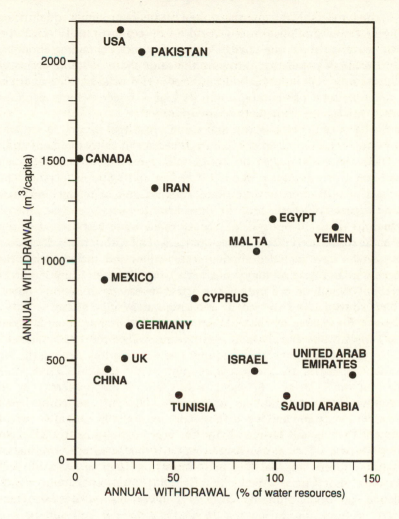

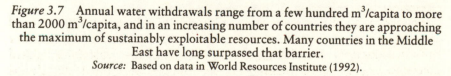

Figure 3.7 Annual water withdrawals range from a few hundred m³/capita to more than 2000 m³/capita, and in an increasing number of countries they are approaching the maximum of sustainably exploitable resources. Many countries in the Middle East have long surpassed that barrier.
Source: Based on data in World Resources Institute (1992).

China illustrates best the regional and local magnitude of the problem. Of the 6 Tt (trillion tonnes) of precipitation falling annually on the country, about 45 per cent end up as stream runoff. About 40 per cent of this flow is potentially usable, and the actual annual withdrawals during the late 1980s amounted to just over 500 Gt, or less than half of the available potential (Smil 1993). Recent per capita use has thus been less than 500 t/year, equal to less than a quarter of the US rate, but roughly equal to some European (Swedish, Polish) uses (World Resources Institute 1992). But the North – covering about one-third of China's territory, containing about two-fifths of China's population, growing the same share of staple grains and producing nearly all industrial output – receives only about one-quarter of the country's total precipitation, and its high summer evapotranspiration leaves it with less than one-tenth of stream runoff.

In the basin of the Huang He, the region's principal stream, less than 20 m^3 of water runoff are available for each hectare of cultivated land and no more than about 600 m^3/person; comparable rates in South China's Chang Jiang basin are, respectively, about 170 m^3/ha and 2,800 m^3/person. These shortages of surface water necessitate a high degree of reliance on underground reserves but here, too, the northern provinces are disadvantaged. While only about 30 per cent of underground water resources are in the North, the region accounts for three-quarters of all withdrawals. Predictable consequences have included sinking water-tables and surface subsidence, especially in and near all major northern cities. And the Chinese estimate that about 50 million peasants in the driest, mostly mountainous, parts of the North do not even have enough drinking water (Smil 1992b).

There is no reliable world-wide estimate of the decline in underground water supplies, but sharply increased pumping since the 1950s has led to many local and regional cases of aquifer overexploitation. Among the most important examples are the recent changes in the Ogallala (High Plains) aquifer in the USA and in the North China Plain aquifers. The first underground storage underlies about 450,000 km^2 (an equivalent of Sweden) in eight states and it provides water for about five Mha, or one-fifth of America's irrigated farmland. By the late 1970s some 170,000 wells lowered the water-level everywhere except in a small area in Nebraska, with the greatest drops approaching and surpassing 10 m in parts of Texas, Oklahoma and Kansas (Weeks *et al.* 1988). Concerns about the fast depletion, increased pumping costs and lower grain prices resulted in reduced withdrawals during the 1980s, but the use still surpasses the recharge by a large margin (Schwarz *et al.* 1990).

Irrigation with stream waters on the North China Plain – a 300,000 km^2 watershed (an equivalent of Italy) of the lower Huang He, Huai He and Hai He – is made difficult by an almost non-existent slope (1:10,000), highly variable summer runoff, and high sediment loads. During the 1950s only about 10 per cent of farmland in the plain's three driest provinces was irrigated – but by the late 1980s the region had more than two million deep tubewells irrigating over 11 Mha of farmland, withdrawing between 25-35 Gt of water a year (Huang 1988), and causing widespread declines of local water-tables.

Even when available, both surface and underground water may not be usable because of unacceptably high levels of contamination by pathogens, industrial effluents (metals, hydrocarbons, polychlorinated biphenyls), agricultural chemicals (nitrates, pesticides) or salts washed from urban areas or deposited from irrigation water. Stream and lake pollution is undoubtedly another prominent world-wide environmental degradation with enormous regional and local consequences. In spite of some notable recent improvements – above all the decline of phosphate concentrations in European and North American waters – the world-wide total of untreated waste water releases continues to grow. Proper treatment and reuse of waste water and more efficient methods of irrigation and industrial production can ease many current water shortages, but a number of countries will be coming up against water barriers, approaching the limits of minimum requirements (Falkenmark 1986).

Higher use-efficiencies can move the barrier, but they cannot eliminate it. For example, even if China would, sometime during the next century, withdraw every drop of its 1.1 Tt of usable surface flow, its average per capita water availability would be less than 1000 m^3, or just 40 per cent of actual US per capita withdrawals during the 1980s! Obviously, such a fundamental physical limitation must have a profound influence on the country's development. Water scarcities will also have an increasingly prominent part in political conflicts. Allocation of river flows between Turkey, Syria and Iraq is already a contentious matter, and the Israeli use of the West Bank aquifer presents an enormous obstacle to any permanent peaceful resolution of one of the world's most intractable enmities (Lowi 1992).

Because the annual flow in the Jordan River equals less than one-third of Israel's water demand, most of the country's water-supply comes from two aquifers. The coastal plain formation supplies about 260 Mt a year; the much richer Yarqon-Taninim aquifer, providing about 590 Mt, underlies the western foothills of the West Bank, and its natural recharge flows in a westerly direction. Although only less than 5 per cent of its recharge is located west of the original Israeli-Arab border, the aquifer can be tapped inside Israel. Since the 1967 occupation of the West Bank Israel has been using disproportionately large volumes of its water, but this disparity, now a matter of critical existential need for Israel, could hardly continue if there would be an independent Palestinian state.

In terms of affected populations, the near-term outlook is certainly worst in Africa, where massive water scarcity is now threatening countries with about two-thirds of the continent's people – yet very little is being done to deal with this change (Falkenmark 1989). The worst affected countries include not only the whole Arab North, but also all Eastern Africa except for Tanzania and Uganda, as well as Lesotho and Namibia in the South and the Sahelian belt (Mauritania to Niger) in the West. Other populous regions rapidly approaching their water barriers are parts of Central Asia and north-western China. Irrigation along Amu-Dar'ya and Sy-Drar'ya has

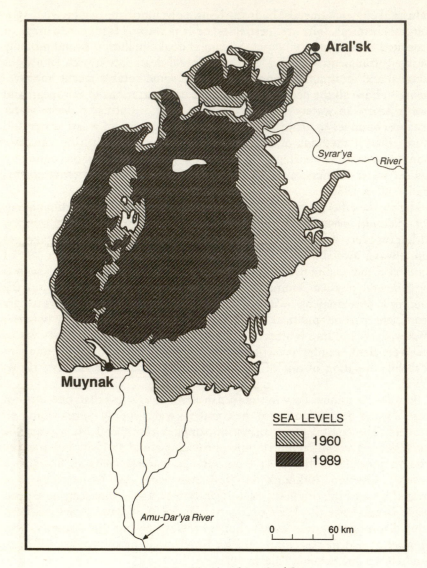

Figure 3.8 The shrinking Aral Sea.
Source: According to maps in Frederick (1991) and Precoda (1991).

become the economic mainstay of Uzbekistan – but it has caused now probably irreversible desiccation of the Aral Sea (Figure 3.8). Formerly the world's fourth largest inland water body, the Aral Sea has lost 60 per cent of its volume, and dropped by 12 m, since 1960 (Frederick 1991; Precoda 1991).

LOSS OF BIODIVERSITY

It may be every ant we trample on
is single before God, Who counts on it
for the unfolding of the measured laws
which regulate His curious universe.
The entire system, if it was not so,
would be an error and a weighty chaos.

Jorge Luis Borges, *Poem of Quantity* (1957)

Not for the first time I find a stanza of Borges's poetry more persuasive than pages of scientific writings. In this case his imagery captures the profound ethical consequences of human evolution and civilizational advances. Our existence and our desire for higher quality of life inevitably threatens the survival of numerous species standing in our way. Until recently our trampling has been done with abandon, only marginally tempered by moral appeals against cruelty to animals and aesthetic arguments for preservation of some spectacular landscapes. Now our deeper understanding of life's richness, our awe at the enormity of genetic information amassed by evolution in millions of species, our realization of the possibility of very rapid losses of a large share of this diversity, and our feelings of moral obligation as well as pragmatic economic and social interest are forcing us to decide how much trampling we can tolerate.

These decisions are immensely complicated. They must consider the conflicts between abstract ethical imperatives and entrenched economic behaviour, and between arguments about potentially invaluable future losses (advanced largely by scientists of rich nations) and immediate existential gains (accruing to some of the poorest people of poor countries). Fundamental scientific difficulty with preservation of biodiversity is that we simply do not know which and how many species we can lose without compromising the sustainability of affected ecosystems, or, with continuing large-scale losses, without undermining the biotic foundations of our societies and economies.

Concerns about the loss of biodiversity are very diffuse, ranging from traditional worries of wildlife conservationists trying to preserve large mammals, to more recent fears about a total destruction of tropical rain forests portrayed as an enormous storehouse of fabulous genetic riches hiding such invaluable treasures as countless cures of major civilizational diseases, effective biodegradable pesticides, and high-yielding varieties of new food crops. Naturally, neither the loss of genetic variability among individuals of a single species nor the reduction of species variety is desirable. The first kind of loss puts many species on an inexorable path to extinction, the other change may have profound, and often irreversibly destructive, effects on the functioning of the whole community of organisms.

But deploring the senseless loss of biodiversity – be it for ephemeral cropping of fields established by burning a patch of tropical trees or for a subdivision with a boat access causing total destruction of a rich temperate wetland – and arguing in favour of effective species conservation does not appear to me to be in the same category as the recent wave of doom-laden writings. This line of arguments, started by Myers (1979) and carried on by Ehrlich and Wilson (Ehrlich and Ehrlich 1981; Wilson 1986; Ehrlich and Wilson 1991) foresees no other outcome but an unprecedented spell of megaextinction (with millions of species destroyed in less than a single generation) entailing an immeasurable loss of genetic information and an unquantifiable but undoubtedly severe price for human survival. Such views deserve a critical look.

Biodiversity of the Earth is still known only very poorly. The best compendia list about 1.4 million living species distributed highly unequally among the five kingdoms (Parker 1982). Monera, including all bacteria and cyanophyta (formerly known as blue-green algae) add up to less than 5000 species, Fungi to ten times as much; there are only about 30,000 species of Protoctista, almost 300,000 species of Plantae (including algae), and just over one million species belong to the kingdom of Animalia (Barnes 1989). But less than 45,000 of that large total are Chordata, with birds adding up to around 9000 and mammals to just 4000 species, compared to some 750,000 insect species and 125,000 other arthropods (Figure 3.9).

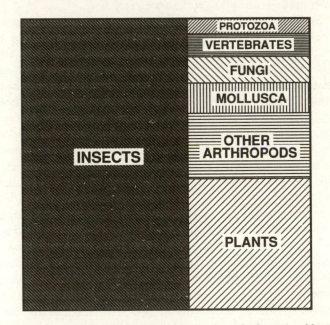

Figure 3.9 Composition of the known biota is heavily dominated by insects; vertebrates are relatively insignificant in terms of total species.
Source: Based on data in Barnes (1989).

Because of major uncertainties concerning the number of tropical insects (their potential count escalated with sampling from previously inaccessible tree canopies), the total number of distinct species may be at least three times and perhaps as much as twenty times as large (Erwin 1986), although a critical appraisal puts the most likely maximum at around five million (Gaston 1991). The new total would only increase the dominance of arthropods. This reality must be kept in mind when trying to assess the consequences of substantial species loss during the coming generations. This demise has been repeatedly labelled as the greatest biodiversity decline since the late Cretaceous extinctions some 65 million years ago – but one proceeding at an incomparably faster rate (decades rather than millions of years).

There are several problems with near-apocalyptic stories of megaextinction. First, all estimates of future extinction rates are mere guesses differing by an order of magnitude and ranging from as little as 2 per cent (Ehrlich and Wilson 1991) to as much as 25 per cent by the year 2000 (Myers 1979). With the total number of species a matter of yet another wide-ranging guess, it makes little sense to give any relative numbers. Second, absolute estimates, most prominently Wilson's 'hundred thousand a year' are not based on any systematic long-term survey of representative ecosystems, but merely on educated guesses building primarily on a number of anecdotal findings (Mann 1991).

Third, predictions of species loss have been based on two dubious assumptions: by applying an island extinction model to large terrestrial ecosystems, and by claiming that an infinite increase in area implies an infinite increase in biodiversity. The foremost weakness of this reasoning is the fact that many tropical rain forests are not at all rich in tree species, and hence losses of substantial areas of such communities would not entail major declines in biodiversity. Forests dominated by a single tree species occur fairly commonly in South-east Asia among Dipterocarps, and in Africa in the north-eastern basin of the Congo River (Connell and Lowman 1989). In Amazonia a tree-seized babassu palm forms virtually pure stands covering nearly 200,000 km^2, mainly in Maranhão, Goiás and Rondônia (Anderson et al. 1991).

Moreover, many contiguous forests have been reduced to a complex pattern of 'islands' – virtually all of Europe and the North American continent east of the Mississippi – and this change, while reducing the diversity of some ecosystems, has not been accompanied by megaextinctions. Even if more vulnerable, large 'islands' of tropical rain forest should retain much of their biodiversity. And, as seen in Puerto Rico, it is possible that there will be substantial restoration of tree cover in deforested areas (Lugo 1986). Massive deforestation did not result in correspondingly large species extinction as the island's secondary mountaintop forests acted as *refugia* for animals and as a base for the gradual re-establishment of tree cover.

Secondary forests, now accounting for about two-fifths of all wooded tropical areas, are rapid accumulators of biomass and nutrients, and if properly managed they can become an important resource (Brown and Lugo 1990). There has also been significant expansion of natural forest into grasslands in the Ivory Coast and in different parts of Brazil (Proctor 1991). Possibilities of forest recovery are also strengthened because of the growing realization of the complexity of tropical soils. Long-standing perception has been one of poor soils low in organic matter which, once deprived of their tree cover containing rapidly cycling nutrients, are rapidly weathered, irreversibly hardened and become unable to support any highly productive growth for centuries.

Evidence that these lateritic soils (latosols or oxisols) are not so dominant as generally believed has been accumulating for decades, but only the recent advances in mapping the Amazonia give the best idea of prevailing misconceptions (Richter and Babbar 1991). The latest maps reduce the area of highly weathered oxisols from about 67 to just 39 per cent, and the extrapolation of these results to other tropical regions suggests that the world-wide area covered by oxisols is most likely around 580 Mha, rather than 1–1.1 Gha still commonly quoted in the ecological literature of the 1980s.

Transferring MacArthur and Wilson's (1963) rule for island biodiversity (number of species being proportional to area$^{0.27}$) to tropical deforestation ignores the fact that the species–area curve levels off within a single biotic community. As a result, in extensive areas of a fairly uniform forest, even a substantial habitat destruction may mean hardly any loss of species. Naturally, destruction of localized, unique, isolated communities would entail major biodiversity cuts. Yet another overlooked reality is the enormous resilience of many ecosystems. Conventional ecological ideas about living systems moving through successional stages to a climax – a steady state of high complexity guaranteeing stability – have had a profound influence on public perception of ecosystems. They are generally seen as fragile (no single adjective has probably been used more often in popular writings) and their disruptions are believed to be largely irreversible.

But this view of Eden-like balances marred by human depredations is just wishful thinking, partly a misinterpretation of complex evidence, partly an idealizing desire to see natural models of perfection contrasting with human depravity. On the most general level, Curry (1977) pointed out an obvious fallacy of this preoccupation with stability:

> The development of man . . . has now provided the noetic impetus to control ecosystem balances. We perceive the maintenance of ecosystem stability and diversity as necessary for the survival of the earth's present component of organisms, including ourselves. In a way, this is antievolutionary.

Indeed, it is.

In the words of Robert Colwell's (1985) summary on the evolution of ecology,

> the possibility that unique, episodic, and catastrophic events have deeply marked the structure of communities, the composition of biota, and important features of organisms and their gene pools is now widely acknowledged in both ecology and evolutionary biology.

These disturbances – ranging from such small-scale, rapid events as storm-induced treefalls and wave damage to tidal pools, to extensive and prolonged impacts created by continental glaciation and changing sea level – generate spatial mosaics of different-aged patches thus permitting the continuation of species with excellent pioneering but poor competitive abilities and they guide adaptation of individuals and systems in evolutionary time.

This view necessitates the redefinition of stability. A traditional way of looking at the stability of ecosystems would be to monitor the degree of constancy in the number of organisms – while a more realistic way is to search for the persistence of relationships. In the first case the concern is with the maintenance of equilibrium in the face of disruptions, and terms such as constancy, endurance, resistance, resilience and inertia have been used by ecologists to characterize a system's behaviour. The second approach does not look for maintenance of equilibria: rather, it focuses on the very ability to survive interferences and then to recolonize the previously inhabited space.

Connell and Sousa (1983) labelled these two viewpoints stability and persistence, and examined a large number of long-term population studies spanning at least one complete turnover of all individuals in order to distinguish a subset of populations or communities that exist in an equilibrium state. Their analysis revealed a continuum of temporal variability with no clear demarcation between populations and communities existing in equilibrium state and those that do not, with only a few examples of truly stable limit cycles, and with no evidence of multiple stable states. Connell and Sousa's conclusions are that 'if a balance of nature exists, it has proved exceedingly difficult to demonstrate' and, consequently, 'rather than the physicist's classical ideas of stability, the concept of persistence within stochastically defined bounds is, in our opinion, more applicable to real ecological systems'.

Careful examination of the natural history of forests illustrates this flexible paradigm quite convincingly. For example, Raup (1981b) found in Massachusetts that destruction of whole stands of native trees by the 1938 hurricane was just the latest instalment of repeatedly severe damages which can be dated to 1815, 1635 and to the latter half of the fifteenth century. Raup's conclusion: 'At least in our part of New England it is probable that no major forest tree has even lived out its possible life-span'. Plants have been adapting to severe disturbances by the evolution of great diversity not only among the species but also among the populations of individual species. These disturbances have been so frequent, and so widespread, that no steady state may have ever existed. And so the ideas (ideals) of orderly succession and admirable stability must be discarded in favour of persistence demanding continuous adjustments to changing environment.

There is also no doubt that the concerns about biodiversity losses are translated into public, and often into policy-making attention, in highly biased ways. Tropical rain forests certainly contain a large variety of potentially useful organic compounds, but repeated portrayal of their plants as fabulous storages of new drugs almost certainly overstates the case (Keay 1990). Mammals (grizzly bears, wolves, elephants, cheetahs) and birds (California condor, whooping crane, grey owl) receive disproportionate attention, as do exotic (to inhabitants of rich northern hemisphere nations) ecosystems in general, and the Amazonian tropical rain forest in particular.

Some recent American efforts devoted to saving endangered species look especially irrational. When a Florida panther is hit by a car, a veterinarian, a tracker with a dog, and two wildlife biologists are dispatched in two helicopters to the site of the accident, the animal is brought to a zoo hospital, treated and repeatedly operated on, and released after more than six weeks of intensive care (Osofsky 1988). In 1990 Riverside County in California spent US $20 million to buy land for the preservation of Stephen's kangaroo rat – and if not rejected by voters it would have budgeted another US $100 million for the rescue (Mann and Plummer 1992).

But a great deal of concern about future species loss must remain even after pointing to a number of questionable assumptions and unjustifiable exaggerations. Concerns about species loss will always be rooted, not only in scientific arguments regarding the benefits of biodiversity for functioning of natural ecosystems as well for the high productivity of farming – but also in often powerful emotions. Effective policies will have to channel these emotions away from the focus on a small number of mammals and birds to the preservation or rational management of total ecosystems, be they tropical, temperate or boreal.

RISKS TO ATMOSPHERIC INTEGRITY

Now clouds will gather when, as on they fly
Through these high realms of heaven, in numbers vast
Bodies of rougher mould will meet, so formed
That though in slender wise linked each on each,
Yet they can cling in unison.

Lucretius, *De rerum natura*

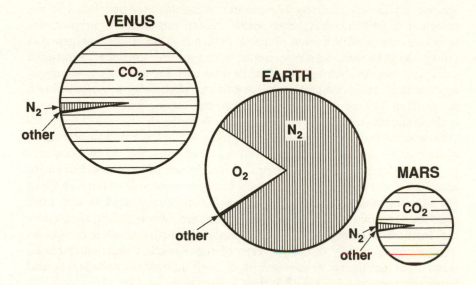

Figure 3.10 Earth's unique atmosphere in comparison with the two very similar compositions on its two planetary neighbours.
Source: Drawn from data in Weast (1990).

There would be no biosphere without the Earth's peculiar atmosphere. Its uniqueness is obvious in comparison with the Earth's two planetary neighbours, Venus and Mars (Figure 3.10). Their atmospheres are made up overwhelmingly of carbon dioxide: Venus has 96 per cent of CO_2, 3.5 per cent N and traces of noble gases, water and ozone; Mars has 95.3 per cent CO_2, 2.7 per cent N, 1.6 per cent Ar and also traces of water and ozone (Weast 1990). Similarly composed, Earth atmosphere would keep the average ground temperature at more than 200°C and the surface pressure would be a few MPa (megapascal). Such conditions would be absolutely inimical to the presence of living organisms made up of carbonaceous compounds assembled in wet tissues.

In contrast to the other terrestrial planets, CO_2 makes up less than 0.04 per cent of the Earth's atmosphere which is dominated by molecular nitrogen (78 per cent) and oxygen (21 per cent). There are also traces of water vapour and noble gases, as well as the unexpected presence of CH_4, N_2O and NH_3 whose concentrations, although relatively very low, violate the rules of equilibrium chemistry to an infinite extent (Lovelock 1979). This atmosphere maintains average planetary ground temperature at 16°C and the surface pressure just around 101 kPa (kilopascal), conditions guaranteeing the continuous presence of liquid water and thermal environment suitable for a wide range of complex life forms.

Such a remarkable match is no accident: the Earth's unique atmosphere is largely a creation of planetary life. Atmospheric composition has changed appreciably during the long coevolution of atmosphere and life, but the principal constituents have been stable during the past 70 million years which have seen the diffusion of mammals culminating in the emergence of our species. And there is little we can do to change the atmosphere's nitrogen or oxygen content. Nitrogen used as the feedstock for synthesis of ammonia represents a totally insignificant share of the element's vast tropospheric stores (Smil 1991b) – and denitrification, the closing arm of the nitrogen cycle, eventually returns all of these molecules to the air.

Oxygen is removed by combustion of fuels. Complete combustion of 1 kg of carbon in solid fuels consumes 2.67 kg of oxygen, and the burning of 1 kg of hydrocarbons needs 4 kg of O_2. Recent world-wide combustion of about 280 EJ (exajoules) of fossil fuels has been removing around 18 Gt of O_2 annually, and combustion of all fossil fuels between 1850 and 1990 consumed roughly 500 Pg (petagrams) of oxygen. When adding O_2 used by the biomass combustion, the total annual anthropogenic reduction of oxygen equals merely 0.002 per cent of the element's huge atmospheric mass. Even a complete exhaustion of all coal, oil and gas resources would lower the tropospheric O_2 by less than 2 per cent.

Observations during the twentieth century have not shown any appreciable shift from 20.95 per cent O_2 by volume, and our improved understanding of the complexity of feedbacks controlling the element's atmospheric presence (Holland 1985) makes us confident that a marginal depletion will be of no consequence for a long-term prosperity of fossil-fuelled civilization. Consequently, anthropogenic risks to atmospheric integrity do not come from interference with the two main constituents, but rather with changing concentrations of trace compounds. World-wide changes include the increased generation of particulate matter by processes ranging from combustion to excessive soil erosion, acid deposition from combustion of sulphur-rich fuels, photochemical smog resulting from complex atmospheric reactions of compounds emitted from the combustion of hydrocarbons, and concentrated heat releases from major urban and industrial areas.

The global ones fall into two distinct categories. The first one involves a gradual but, in evolutionary terms, relatively rapid alteration of the Earth's radiation balance brought by rising atmospheric concentrations of trace gases whose molecules absorb part of outgoing infra-red radiation. For many decades this process was seen almost solely as a matter of higher CO_2 levels caused by human interference in the carbon cycle: by combustion of fossil and biomass fuels and by conversion of natural ecosystems. But accumulation of other 'greenhouse' gases – most notably CH_4 and chloro-fluorocarbons (CFCs) – is already as important as higher CO_2 levels. The second global risk arises from the effect of CFCs on the stability of stratospheric ozone.

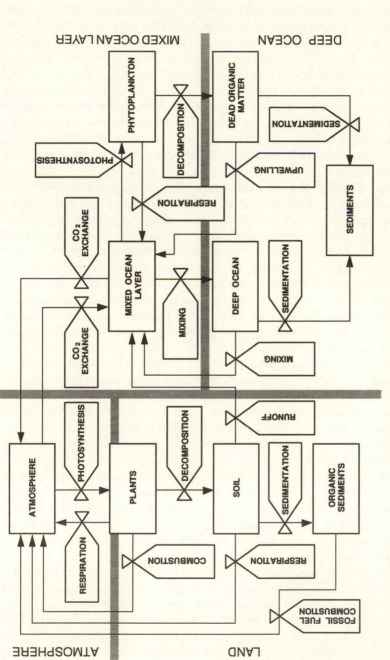

Figure 3.11 Biospheric carbon cycle circulates the element rather rapidly between the atmosphere, biota and the mixed ocean layer. In contrast, the exchange between that layer and the deep ocean is very slow.
Source: Based on Smil (1985a).

Greenhouse gases and the risk of planetary warming

CO_2 is the most important greenhouse gas and the understanding of its atmospheric dynamics must start with the global carbon cycle. Carbon, accounting for almost half of all biomass, forms the basic matrix of life. The element is doubly mobile, as it cycles both through waters and the atmosphere where it is present overwhelmingly as CO_2 (Figure 3.11). The gas is not only the source of carbon for photosynthesis, but also an essential regulator of planetary temperature: without it the Earth's surface would be about 33°C colder and unable to support life. Reasons for this importance are well understood. The gas is essentially transparent to the incoming, short-wave solar radiation while being a vigorous absorber of the outgoing, long-wave terrestrial radiation which it then emits both to space and back toward the surface.

Analysis of ice-trapped air bubbles shows that over the past 160,000 years CO_2 concentrations fluctuated between 150 and 300 ppm, with marked highs during interglacials and lows during glacial periods (Boden *et al.* 1990). Since the middle of the nineteenth century the slow natural fluctuations of atmospheric CO_2 have been overwhelmed by the exponential growth of fossil fuel combustion releasing carbon previously sequestered for tens to hundreds of million years in coals, oils and gases, and by the concurrent extensive conversion of grasslands and forests to farmland and built-up areas. CO_2 levels rose from 275–80 ppm during the eighteenth century to about 300 ppm by the beginning of the twentieth century, surpassed 310 ppm by the early 1950s (Figure 3.12).

Systematic monitoring of background CO_2 levels in remote places started at the Mauna Loa Observatory on the island of Hawaii and at the South Pole in 1958. By 1990 the annual mean surpassed 350 ppm. Annual increases during the 1980s fluctuated between 1.1 and 1.8 ppm, or roughly 0.3–0.5 per cent of the total level. Compared to the 1850 average of about 280 ppm, the tropospheric CO_2 level rose by about 25 per cent during the age of industrialization (Figure 3.12). Annual releases from combustion of fossil fuels have been recently about 20 Gt. Natural gas flaring, cement production and waste combustion are smaller sources whose total is less than the uncertainty surrounding the net CO_2 emissions from tropical deforestation and conversions of grasslands and wetlands to fields. Recent estimates for total emissions from these sources are between 2 and 9 billion t of CO_2 (Houghton *et al.* 1990).

The best estimate for cumulative 1850–1990 emissions from fossil fuels is roughly 700 Gt. CO_2 releases from ecosystemic changes caused by pioneer agriculture and deforestation are much more difficult to estimate but aggregates close to 600 Gt are most plausible. The grand total of all anthropogenic CO_2 releases for the period 1850–1990 is then about 1.3 Tt. If all of this gas would have stayed in the atmosphere the tropospheric levels

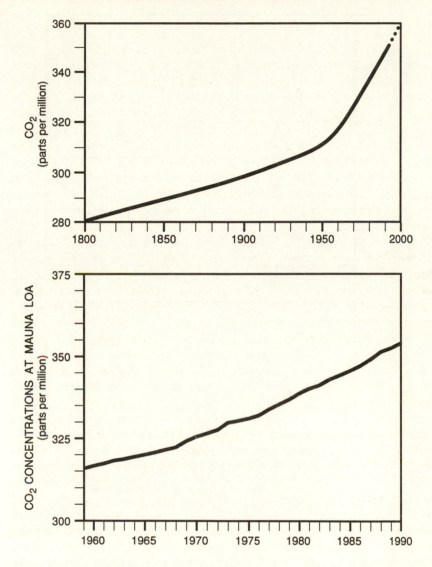

Figure 3.12 Approximate trend of global atmospheric CO_2 since 1800, and annual averages at the Mauna Loa observatory, where the concentrations surpassed 350 ppm by 1989.

Sources: Plotted from annual averages in Smil (1985a) and Ferguson (1991).

would have increased by nearly 170 ppm; in reality, only about 40 per cent of the post-1850 anthropogenic CO_2 input has been retained in the atmosphere. The rest has been sequestered in the oceans, or absorbed by vegetation.

Until the early 1980s increasing CO_2 concentrations were the sole focus

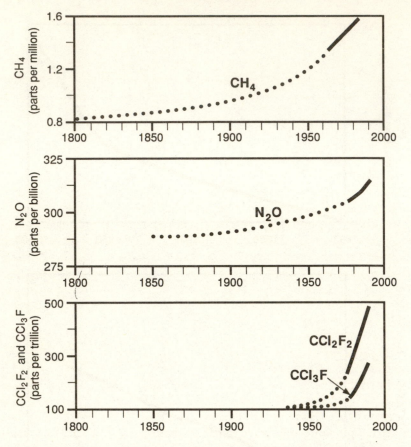

Figure 3.13 Increasing atmospheric concentrations of four other leading greenhouse gases: methane, nitrous oxide and two CFCs (dotted lines indicate reconstructed trends, solid lines are actual measurements).
Sources: Based on Stauffer (1985), Pearman *et al.* (1986) and Ferguson (1991).

of studies and computer models of changing atmospheric radiation balance. But atmosphere contains other trace gases with similar, or even stronger, 'greenhouse' effect on the Earth's radiation budget, and human actions have been increasing concentrations of two naturally occurring compounds – methane and nitrous oxide – and of a group of synthetic CFCs at rates far surpassing the generation of CO_2 (Figure 3.13). CH_4 levels roughly doubled during the industrial era, from about 0.8 ppm at the beginning of the nineteenth century to about 1.7 ppm in the late 1980s (Boden *et al.* 1990). And since its measurement started in 1977 N_2O concentration rose from about 300 ppt to around 310 ppt by 1989, while those of chlorofluorocarbons have roughly doubled during the same period, with CFC–12 (dichlorofluoromethane) approaching 500 ppt and CFC–11 (trichlorofluoromethane) close to 300 ppt (Thompson *et al.* 1990).

Anthropogenic sources of CH_4 include anaerobic fermentation of solid wastes in landfills and garbage pits and of organic matter in flooded farm soils, enteric generation of the gas by livestock, and the direct emissions from coal-mines and natural gas wells and pipelines. Estimates for all of these categories have large margins of uncertainty, with global totals ranging from less than 200 to just over 500 Mt CH_4/year (Watson *et al.* 1990). The most detailed disaggregated summary puts the global anthropogenic CH_4 flux at about 255 Mt in 1987, with the US contributing nearly 17 per cent and China in the second place with more than 11 per cent of the total (World Resources Institute 1990).

Principal anthropogenic sources of N_2O are intensifying fertilization and biomass and fossil fuel combustion. Fertilizer nitrogen is released as N_2O mostly through denitrification, the essential closing arm of the element's biogeochemical cycle (Golerman 1986). Extensive measurements of denitrification reveal enormous differences in diurnal and seasonal rates (McKenney *et al.* 1978). Consequently, the global estimates of fertilizer-derived N_2O emissions range from a negligible total of about 15,000 t to as much as 3.5 Mt (Watson *et al.* 1990). Biomass and fossil fuel combustion may raise the total to a maximum of 4.2 Mt N_2O.

CFCs are synthetic compounds used since the 1930s as refrigerants, aerosol propellants, foaming agents and cleaners, and they will be discussed in greater detail in the next section. Their production is heavily concentrated in industrialized nations, with the US contributing about 25 per cent (World Resources Institute 1990). CH_4, N_2O and CFCs are much more efficient absorbers of long-wave radiation than CO_2. One kg of methane will absorb about 70 times more long-wave radiation than a kg of CO_2, the CO_2 heating equivalent for N_2O is about 250, and for CFCs it is of the order of 10,000 (World Resources Institute 1990). Consequently, for an estimate of relative net heating effect we must first multiply the emissions rates by average retention fractions, and then by CO_2 heating equivalents. In 1990 CO_2 accounted for slightly less than 60 per cent of the total effect, CFCs for roughly a quarter and CH_4 for one-sixth.

In spite of decades of scientific interest in the possibility of global warming we have no reliable understanding of biospheric consequences which could result from continuing accumulation of greenhouse gases. Our forecasts are largely speculative, our models are still too simplistic. Palaeoclimatic analogies, better understanding of atmospheric physics and availability of more complex computer models narrow the ranges of possible outcomes – but major uncertainties will not be removed soon. The most-often-cited conclusions are the results of American computer models predicting that the doubling of pre-industrial CO_2 (that is, concentration of about 600 ppm or its equivalent composed of a mixture of greenhouse gases) will raise tropospheric temperatures by anywhere between 1 and 5°C, and that this

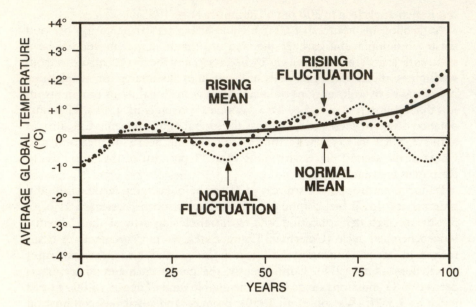

Figure 3.14 Difficulties in detecting a clear warming signal are illustrated by a hypothetical warming scenario: it can take decades to distinguish unequivocally the normal climatic fluctuation from the fluctuating temperature rise.
Source: Based on Mahlman (1989).

warming will be about two to three times more pronounced in higher latitudes than in the tropics and greater in the Arctic than in the Antarctic (Houghton *et al.* 1990).

This warming should bring intensification of the global water cycle accompanied by major redistribution of precipitation patterns, intensification of extreme weather conditions (tropical and temperate cyclones, droughts), shifts in ecosystem boundaries (northward march of grasslands and forests) and changes in plant productivity, poleward retreat of sea ice, and appreciable sea-level rise (causing coastal inundation, greater erosion and storm damage, and greater salt water intrusion into coastal wetlands and river estuaries). None of these general qualitative conclusions can be translated into a firm quantitative prediction.

Nor is there much confidence in ascertaining the degree of past warming. According to general circulation models a 25 per cent increase of atmospheric CO_2 should have caused global warming of between 0.4 and 1.1°C. The two most extensive studies of the existing global record of surface temperatures (Jones *et al.* 1986; Hansen and Lebedeff 1988) detected long-term planetary warming of, respectively, 0.5 and 0.78°C. But the uncertainties of this record, above all the changes in measurement techniques, station location and station environment, make it impossible to claim unequivocally that all, or even most, of this warming trend has been real.

There are many plausible explanations of why the warming may have been smaller than the one predicted for a 25 per cent increase of CO_2: the slight warming trend may be an entirely natural change; it may be a manifestation of the greenhouse effect but one delayed by ocean heat storage; it may have been attenuated by increased oceanic or anthropogenic production of sulphate aerosols which tend to reduce surface temperatures – or even partially counteracted by the onset of a new global cooling. In any case, detection of a clear warming signal may be delayed for decades (Figure 3.14). As there is no way to eliminate easily both the prospective (inherent weaknesses of general circulation models) and retrospective (quality of climatic data) uncertainties, the coming decades will see a continuing clash of claims regarding the presence and extent of the CO_2-induced warming, detection of sure signs of such a cause-effect link, and the degree of future biospheric impact.

While doubting and criticizing various conjectures concerning the consequences of future greenhouse warming is imperative, dismissing them outright would be foolish. And so, acting as prudent risk minimizers, we should be aware of what may be the most unwelcome consequences of such a global change – as well as some likely benefits. A rapid, catastrophic sea-level rise is the most popularized risk – but not the most likely one. The sea-level has been rising during the past 100 years but virtually all of this gentle increase is due to the retreat of North American glaciers and to the wastage of relatively small Central Asian and Patagonian ice caps. There has been no decay of the vast Antarctic Ice Sheet. Indeed, that huge mass of ice is most likely growing (*Ad hoc* Committee on the Relationship between Land Ice and Sea Level 1985).

The best glaciological consensus estimates the contributions to mean sea-level rise by ice wastage caused by the doubling of pre-industrial atmospheric CO_2 level at between 10–30 cm for glaciers and small ice caps, the same amount for the Greenland Ice Sheet, and 0 to 30 cm for the Antarctic ice. Consequently, the total sea-level change attributable to CO_2 doubling could be as small as 20 cm or as high as nearly one m. Whatever the actual mean rise, there would be significant local departures (Bryant 1987). A rapid and total melting of the West Antarctic Ice Sheet that could conceivably result in a 4–5 m sea-level rise within a century appears to be extremely unlikely.

But even a relatively small but fairly rapid sea-level rise could be costly in both ecosystemic and economic terms (Broadus 1989). Not all effects would be negative. Higher water temperatures and higher influx of nutrients into coastal waters could enhance photosynthesis and promote the growth and survival rate of ocean fish (Sibley and Strickland 1985) – but such gains may be totally negated by the loss of wetland habitats and the costs of shoreline protection. Even a sea rise limited to less than one metre would have catastrophic effects on such densely populated low-lying lands as the Nile

97

Delta or Bangladesh (Milliman *et al*. 1989).

Planetary warming would eventually transform also many ecosystems. Palaeoclimatic evidence shows the sensitivity of forests to relatively small temperature changes: a rapid warming could preclude effective adaptation and the southernmost forest boundaries could recede northward (Solomon and West 1985). A long transitional period – when the dying forests and their soils would release more CO_2 and CH_4 into the atmosphere – could be a time of a great ecosystemic upheaval and economic hardships. Warming of boreal ecosystems could also accelerate the disintegration of organic matter in bogs and tundras, again an additional source of CO_2 and CH_4. And reduction of precipitation and soil moisture in major agricultural regions, and substantial changes in river runoff could have a dramatic effect on the economies of affected areas.

Predictions of ecosystem responses to higher tropospheric CO_2 concentration remain speculative, especially as far as the long-term growth of forests is concerned (Eamus and Jarvis 1989; Mooney *et al*. 1991). Experiments evaluating the short-term (weeks to months) response of seedlings and saplings show enhanced growth of about 40 per cent with doubled CO_2 levels, but these findings cannot be simply extrapolated to long-term exposures of mature trees. No direct information is available to estimate such responses, but the evidence of tree rings points to increased annual increments during the past periods of CO_2 enrichment.

At the same time, it would be sensible to expect that in many cases the initially faster growth will eventually become limited by the availability of nutrients. These limitations will also largely decide the response of other ecosystems. Open-top chamber experiments showed that doubled CO_2 levels did little for the productivity of a tundra grassland – but resulted in larger and denser growth of marsh sedges (Bazzaz 1990). By far the best response should come from crops whose productivity would not be limited by inadequate nutrient supply (Kimball 1983; Lemon 1983; Enoch and Kimball 1986). There will be considerable species-specific differences, but doubling of pre-industrial CO_2 levels would increase average crop yields by about 30 per cent. The response would be strongest among the C_3 species (wheat, rice) which include 12 out of the world's 15 leading crops, including all staple grains except for corn and sorghum.

Benefits would go beyond high productivity. Doubled CO_2 concentrations would reduce water losses during photosynthesis by about 30 per cent, but in this case C_4 plants (corn, sugar-cane) would respond better. Other important benefits could include lowered photorespiration (daytime oxidation, and hence the loss, of plant mass produced by photosynthesis), higher symbiotic fixation of nitrogen in legumes, increased resistance to lower temperatures and to air pollutants, and a better tolerance of soil and water salinity. Combination of these responses would mean that all well-fertilized crops would yield more in their current environments while using less water and, when rotated with legumes, less nitrogen – or that they could be grown in areas which are now too arid for continuous farming, or that they may be still able to outperform the current yields in those regions where precipitation may decline as a result of planetary warming.

Slightly higher summer temperatures would not present any serious risk to healthy people. *Homo sapiens* has thermoregulatory capabilities unmatched by any other mammalian species (Hanna and Brown 1983). In contrast, higher frequency and longer duration of heat-waves would be a clear concern: copious research documents excessive mortality, above all among elderly people with cardiovascular and respiratory diseases, during heat-waves lasting more than three days (White and Hertz-Picciotto 1985). Warming may also pose greater health risks because of its effects on asthmatics: it could substantially increase the seasonal burdens of temperature- and humidity-dependent allergens.

Perhaps the most worrisome health effect of planetary warming would be the extratropical radiations of tropical diseases promoted by higher temperatures and humidities, longer breeding seasons, and shifts of ecosystem boundaries. Insect vectors and parasites transmitting and causing malaria, trypanosomiasis, leishmaniasis, amoebiasis and filariasis could then spread beyond their current boundaries. In higher latitudes of North America there may be increased risks of Lyme disease, Rocky Mountain spotted fever, dengue fever, and arbovirus encephalitis.

Challenges inherent in understanding the complex genesis and dynamics of climatic change, and in assessing the no less complex ecosystemic, economic and social consequences of higher tropospheric temperatures, mean that the high level of research activities and political interest established during the 1980s will continue for decades to come. Neither the new findings relating global warming to the length, and hence the intensity, of the solar cycle, rather than to anthropogenic emissions (Friis-Christensen and Lassen 1991), nor the indications that the overall warming may be much lower than predicted by standard climatic models (Mitchell *et al*. 1989), will weaken the concerns.

I agree with Broecker's lament that as funding for greenhouse studies has become increasingly available, 'too many scientists with too little to contribute to predicting the fate of the planet' have decided 'to hitch their wagons to the greenhouse star' (Kerr 1991). These armies of instant greenhouse warming experts and the high media interest in a story with a potentially catastrophic outcome will have an obvious influence on the policy-making process – and will make it actually more difficult to come up with effective long-term responses. In comparison, consensus has been much easier in the case of anthropogenic destruction of stratospheric ozone.

Stratospheric ozone and the biosphere

Only a rather narrow band of solar radiation energizes the Earth's life: absorption of light by pigments in chloroplasts has two distinct peaks, one between 420 and 450 nm (nanometers), the other from 630 to 690 nm. Photosynthesis is thus driven largely by a combination of blue and red light. Infra-red radiation helps the process indirectly, by warming the tissues, but the shorter, ultra-violet (UV) wavelengths reduce the photosynthetic rates of many plant species and can damage their cells. Wavelengths between 320 and 400 nm (UVA) are of much less concern than the band between 290 and 320 nm (UVB).

Oceanic phytoplankton, the base of complex marine food webs, appears to be especially sensitive to higher UV levels (El-Sayed 1988). Experiments with a variety of phytoplankton species showed that elimination of both UVA and UVB increased their productivity 2–4 times, while an UVB enhancement caused substantial cuts in the photosynthesis. First extended measurements in Antarctic waters confirmed primary productivity declines of at least 6–12 per cent during springtime ozone depletion (Smith *et al.* 1992). Many higher plants are similarly vulnerable, and some of these effects could translate into reduced crop yields. Even if UV radiation would have no effect on people and animals, any substantial increase in its ground level would have to be a concern.

But excessive UV radiation can damage both humans and animals. There is fairly clear evidence that a 1 per cent increase in UVB flux causes 2 to 3 per cent rise in the incidence of basal and squamous cell carcinomas. An overwhelming majority of these cancers are non-lethal – unlike the malignant melanoma. Although its relationship to higher UVB levels is harder to quantify, there is strong circumstantial evidence that its incidence would also increase. Even small UVB increases would bring higher incidence of cataracts, conjunctivitis, photokeratitis of the cornea and blepharospasm. Preliminary investigations also point to effects on the immune system by damage to suppressor lymphocytes which protect the organism against skin cancer and against bacterial and parasitic invasions through the skin. The most important non-biotic effect of higher UVB flux would be faster degradation of such common plastics as polyvinylchloride.

Ever since the formation of the Earth's atmosphere UVB effects on biota were marginal. Complex terrestrial life developed only after all but a small portion of UVB was prevented from reaching the biosphere. Absorption by ozone molecules in the stratosphere, mostly between 25 and 35 km above the ground, has been the only effective way of protecting the biosphere from an excessive UVB flux. This protective mechanism seemed to be beyond human reach. Systematic measurements of ozone started around the world during the 1950s and for about 20 years there was no indication of any decline. While measurements did not show any significant departure from a

100

random drift, a theoretical prediction caused a major concern about future developments.

Molina and Rowland (1974) concluded that the rapidly growing releases of chlorofluorocarbons (CFCs) would eventually bring their stratospheric concentrations to 10–30 times the mid-1970s level, and that the photo-dissociation of these compounds would liberate so many chlorine atoms that a destruction of an appreciable share of the protective ozone layer would be inevitable. CFCs were formulated during the 1930s to provide better refrigerants. Unlike the previously common compounds (ranging from NH_3 to CO_2) they were stable, non-corrosive, non-flammable, non-toxic – and yet relatively inexpensive. They are also thermodynamically superior, enabling a higher efficiency of refrigeration. After the Second World War CFCs became rapidly dominant in every field of refrigeration, and their desirable properties led to a variety of other uses.

The two most familiar applications have been as the propellants in sprays dispensing aerosols from deodorants to insecticides, and as the refrigerant in air conditioning. More recently a leading industrial use of CFCs has been as foam-blowing agents in producing polyurethane, polystyrene and poly-olefin. Cellular structure of these synthetics with a very wide range of uses (from rigid insulation to fill-in packaging) is achieved by the volatilization of non-reactive CFCs with low boiling point. Other increasingly prominent uses have been as solvents for cleaning electronic circuits, in extraction of edible plant oils and essential aromatic oils, and in degreasing machine parts.

There are many CFCs, but three compounds have been dominant world-wide (Figure 3.15). CCl_3F accounts for about half of all CFC output, CCl_2F_2 for some 30 per cent, and $CHClF_2$ for at least 10 per cent. There is little difference as to their potential to destroy stratospheric ozone. All of them are very long-lived, and they do not appear to have any substantial tropospheric sink. Although much heavier than the air, wind-borne molecules of CFCs randomly diffuse into the stratosphere. Only there they can be dissociated by wavelengths shorter than 220 nm which do not reach lower atmospheric layers. Their breakdown releases free chlorine atoms whose slow downward drift starts an extremely long chain of destructive reactions.

In the absence of CFCs the stratospheric ozone destruction proceeds through two chains of reactions, one involving HO radicals formed from water or methane, the other driven by nitrous oxide from bacterial denitrification. But the chlorine route is vastly more destructive. After a chlorine atom breaks down an ozone molecule and forms chlorine oxide, that compound can react with atomic oxygen and produce an oxygen molecule – and free the chlorine atom to attack another ozone molecule. During one to two years before the chlorine atom eventually reacts with naturally occurring methane and forms a readily water-soluble, and hence precipitable, HCl, it can destroy about 10^5 ozone molecules.

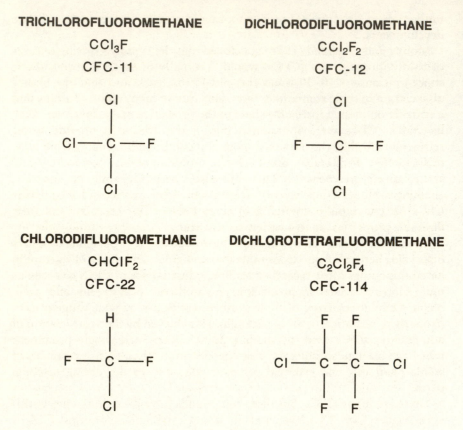

TRICHLOROFLUOROMETHANE

CCl_3F

CFC-11

DICHLORODIFLUOROMETHANE

CCl_2F_2

CFC-12

CHLORODIFLUOROMETHANE

$CHClF_2$

CFC-22

DICHLOROTETRAFLUOROMETHANE

$C_2Cl_2F_4$

CFC-114

Figure 3.15 Principal chlorofluorocarbons (their proper chemical names are followed by their formulae, commercial designations, and structures).

Soon after the Rowland–Molina warnings the US National Research Council initiated a series of modelling exercises, which concluded that the continuation of CFC releases at the 1977 rate would eventually lead to a steady-state reduction of stratospheric ozone by 15–18 per cent (NRC 1979). Then a major downward revision (NRC 1982) concluded that if CFCs releases were to continue at the 1977 rate the steady-state ozone depletion could be just between 5–9 per cent – and in February 1984 another reappraisal (NRC 1984) predicted the most likely reduction to be no more than 2–4 per cent! But easing of concerns about the fate of stratospheric ozone was shortlived. Ozone measurements at the British Antarctic Survey base at Halley Bay showed a slight downward trend between 1957 and the mid-1970s, followed by a rather sharp drop afterwards.

In 1985, after the lowest spring (October) concentrations dropped from about 300 Dobson units (roughly equal to 1 ppbv) to about 200 Dobson units, Farman *et al.* (1985) published the paper which started the intensifying investigation of the Antarctic ozone hole. Since then a series of measurements confirmed the recurrence of major ozone loss above the continent during every spring (Stolarski 1988). The October minimum was slightly higher in 1986, it sank to a record low of less than 150 Dobson units above the South Pole in 1987, recovered to well above 200 Dobson unit during the following springtime, and dipped once again below 200 in 1989. In 1990 the pattern of fluctuating ups and downs was broken as the minima stayed close to the previous year's level (Kerr 1990b). Similarly drastic ozone depletion has not been observed anywhere else, although there seems to be an increasing probability for a major seasonal ozone loss above the Arctic.

Montreal Protocol of September 1987 called for a 20 per cent reduction of global CFC emissions by 1994, and for another 30 per cent cut by the year 2000. In June 1990 the London agreement set the goal of essentially phasing out the production of all CFCs by the year 2000. Detection of unexpectedly high ClO_x concentrations above the mid and high latitudes of the northern hemisphere during the winter of 1991–2 led to calls for accelerated phase-out of all CFCs. But given the long atmospheric lifetime of CFCs, peak tropospheric levels would be reached only a decade after stopping their production, the maximum stratospheric effect will be felt between 2015–20, and the resulting ozone loss will continue to have global implications for the remainder of the twenty-first century (Rowland 1989, 1991). The fundamental, and as yet unanswerable question, is how costly will be the effects of this degradation on human health and on the productivity of various ecosystems.

Other anthropogenic emissions

Urbanization, industrialization, farming and deforestation are obviously large world-wide sources of particulate matter – but effects of these emissions are highly uncertain. To begin with, comparisons of natural and anthropogenic particulate matter are made unreliable by the fact that no large natural contributions – soil erosion, volcanic ash, plant hydrocarbons, and biogenic sulphur and nitrogen compounds – are known with a satisfactory accuracy. Moreover, while emissions from fossil-fuel combustion can be well estimated, quantification of soil erosion caused by farming, and of phytomass combustion is highly uncertain.

These limitations result in a wide range of estimates of human contributions to the world-wide generation of particulate matter. The extremes may be less than 10 and more than 30 per cent. Roughly one half of all anthropogenic particulates comes from fossil fuel combustion. Thanks to extensive controls fly ash from coal-burning has become a minor part of these emissions which are now dominated by sulphates and nitrates generated by oxidation of nearly 200 Mt of SO_2 and 50 Mt of NO and N_2O

103

released by fossil-fuel combustion. Sulphates and nitrates often dominate the mass of fine aerosols between 0.1 and 1 μ, and are responsible for most of the light scattering.

Their greatest impact has been in the northern polar latitudes, where they are the primary constituents of the Arctic haze (Nriagu *et al.* 1991). This thick (up to 3 km), brownish or orange pall covering a circumpolar area as large as North America, is densest during March and April when very low precipitation and very strong thermal inversions limit particle removal. Levels of sulphates in the haze are around $2\mu g/m^3$ (compared to $5-15\ \mu g/m^3$ in non-urban regions of eastern USA and Western Europe) and horizontal visibilities are restricted to 3–8 km. Long-term effects on the region's radiation balance are uncertain.

Sulphate particles are just of the right size to be highly effective at light scattering – but their actual contribution to reduced visibility in (and downwind from) industrialized regions remains uncertain. Extensive US monitoring proved that 40–60 per cent of the fine aerosol mass during both summer and winter haze periods above the eastern half of the country was made up of compounds not associated with either sulphates or nitrates (EPRI 1981). In contrast, subsequent studies in the Ohio River Valley attributed half of all fine-particle mass to sulphates from coal combustion (Shaw 1987).

In any case, sulphates from industrial sources, unlike CO_2, cannot diffuse world-wide because the average residence time for all atmospheric sulphur species is just 20–100 hours, with the mode around 40 hours. Before their wet or dry deposition sulphates will have at most large regional or semi-continental effects on visibility. Yet sulphates can have important regional, and conceivably even global, climatic impact because their absorption of the incoming radiation cools the Earth's surface and warms up the upper atmosphere. While there are still uncertainties about the large-scale cumulative effect of this cooling, there is no doubt that sulphates have been the leading cause of anthropogenic acidification which is affecting extensive areas of Europe and North America and smaller regions of Latin America and East Asia.

Since the early 1970s acid deposition and its consequences has been studied extensively in both Europe and North America (Swedish Ministry of Agriculture 1982; NRC 1983; NRC 1986; Longhurst 1991). By far the most comprehensive set of interdisciplinary studies was produced by decade-long investigations of the US National Acid Precipitation Assessment Program (NAPAP 1990). Anthropogenic emissions of sulphur and nitrogen oxides, generated by combustion of fossil fuels and by several industrial processes, are much more concentrated than natural releases of acidifying compounds. Consequently, downwind precipitation pH may be below its naturally slightly acid level of 5.6 (the consequence of CO_2 forming weak H_2CO_3). Recent anthropogenic sulphur emissions have been comparable to all biogenic releases and may account for as much as 40 per cent of the element's annual global mobilization.

Combustion of fossil fuels, above all of coals, generates about 80 per cent of these releases, with the rest coming mostly from the smelting of non-ferrous metals. Emissions are heavily concentrated in the temperate zone of the northern hemisphere (about 95 per cent between 30 and 55°N), and in the most heavily polluted areas they are as high as 8–12 t SO_2/km^2, much above the biogenic fluxes. Photochemical and wet-phase oxidation of the emitted SO_2 produce sulphates, the dominant aerosols of polluted atmospheres which can cause short-term acidities lower than pH 3 and annual averages in the worst affected areas between pH 4 and 5. Limited tropospheric residence time of sulphates, and the presence of neutralizing alkaline matter over many drier continental areas, prevent a global diffusion of acid precipitation.

Effects are limited to areas about 1500 km downwind from the largest sources. Eastern North America and Western and Central Europe are the most extensive regions with the lowest precipitation acidity (Figures 3.2 and 3.3). Smaller areas are found in Asia, Africa and Latin America (Rodhe 1989). Contrasting views of long-term ecosystemic effects of acidification were put into perspective by the completion of the National Acid Precipitation Assessment Program (NAPAP) in the autumn of 1990. This decade-long effort, involving 2000 US and foreign scientists, confirmed the existence of various impacts, but found their extent and urgency not as serious as many had feared (NAPAP 1990).

Most notably, only about 4 per cent of the lakes and 8 per cent of the upper courses of streams in US areas with elevated acid deposition were actually found to be acidic – and only about half of this degradation can be attributed to atmospheric deposition. And while acidification potentiates the effects of natural stresses effecting tree growth, the high-elevation stands of spruce and fir which have been most affected by acid deposition cover less than 0.1 per cent of the forests in the eastern USA. NAPAP could not confirm any negative effects on agriculture or human health, and it concluded that the prevailing levels of acid deposition are not likely to increase appreciably either the degree or the extent of the damage for several decades.

The verdict – based on the most extensive (and at nearly half a billion US $ also the most expensive) interdisciplinary research ever undertaken to assess an environmental problem – could not be clearer. Acid deposition is no 'death from sky' headlined by many publications in the late 1970s and the early 1980s, but rather a relatively low-level regional problem causing limited acute damage in some sensitive locales – and one which should be readily manageable with existing emission controls. I came to the same conclusions in my assessment of three grand biogeochemical cycles (Smil 1985a).

Emissions causing seasonal or nearly chronic photochemical smog have

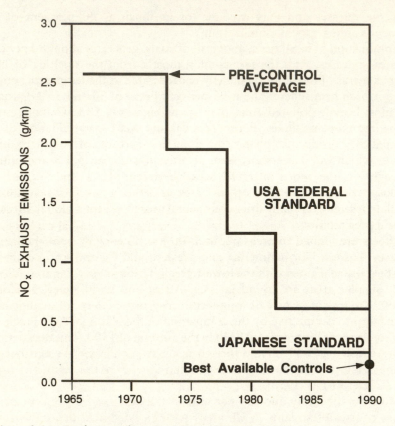

Figure 3.16 Declining exhaust emission of NO_x from American cars compared
with the Japanese standard and with the best available controls.
Source: Based on data in MVMA (1991).

been increasing in virtually all the major urban and industrial areas of every
continent. The process starts with a rapid oxidation of NO to NO_2 involving
a variety of reactive molecules (HO radicals, CO, hydrocarbons, aldehydes).
Subsequent dissociation of NO_2 and oxidization of hydrocarbons leads to
rising concentrations of O_3. Automotive emissions have been by far the
largest source of nitrogen oxides and hydrocarbons, but increasing aircraft
traffic is becoming an important contributor to higher concentrations of
tropospheric ozone (Johnson *et al.* 1992). Highly reactive ozone has a
number of adverse effects on living organisms and on materials: it impairs
lung function, injures cells, limits work and exercise capacity and lowers the
resistance to bacterial infections (Lippmann 1991).

Increasingly effective automotive controls have brought impressive reductions of specific NO_x emissions (Figure 3.16). However, the increasing
concentration of cars in many smog-prone metropolitan areas of the rich
world (Los Angeles, Phoenix, Vancouver, Madrid, Athens, Sydney), and the

106

rapid rise of their ownership and the lack of effective exhaust controls in Asian, African and Latin American cities (Tehran, New Delhi, Bangkok, Taipei, Cairo, São Paulo, Mexico City) mean that urban NO_x emissions are still growing and causing serious photochemical smog.

The wealth of experimental and epidemiological data clearly links automotive emissions with an increase of respiratory diseases in general, and emphysema and asthma in particular, as well as with cancer, heart disease and neurotoxicity, while long-term effects of subthreshold exposures remain uncertain (Watson *et al*. 1988). Effective controls can come only with a combination of still better emission controls, new automotive fuels, alternative transportation modes and limits on the use of private cars.

At least three other potentially dangerous air pollutants can be transported downwind over long distances: dioxin, mercury and lead (Travis and Hester 1991). The first compound is the most potent chemical carcinogen evaluated by the US Environmental Protection Agency, and it is now present, albeit in very low concentrations, throughout the biosphere, including remote Arctic sites. Airborne mercury, released primarily by coal combustion, is clearly a continental pollutant: it contaminates fish in lakes with no nearby waterborne emissions of the metal from paper pulp manufacturing or fungicides.

Airborne lead emissions come from the smelters and from waste combustion, but more than 90 per cent of the world-wide flux originates from gasoline additives. Lead's toxicity to children and its deleterious effects on their mental development have been abundantly documented (Needleman and Bellinger 1991). While the rich world's use of lead in gasolines and paints had rapidly declined during the 1980s, the metal is still widely used throughout the poor world and its current global anthropogenic mobilization is one to two orders of magnitude greater than the release from all natural sources (Hutchinson and Meema 1987).

4

EFFECTIVE STRATEGIES

HOW MUCH CAN WE CHANGE
MODERN CIVILIZATION?

It is, therefore, of the highest importance to gain a clear insight into the
means of modification and coadaptation.

> Charles Darwin, *The Origin of Species* (1859)

Darwin's admonition seems to be no less applicable to assessing the
prospects of modern civilization than it was to tracing the past development
of the biosphere: life's adaptation to environmental changes continues to be
seen as the very essence of evolution. Adaptive processes do one of two
things: they either improve an organism's fit with its environment, or they
improve the efficiency or responsiveness (or both) of its internal functions.
In plants and animals they are realized through natural selection and
mutation, as well as through correlational selection and aggregation of two
or more traits (Endler and McLellan 1988). These processes require usually
very long periods of time and their effects are far beyond the changes
achievable by human societies within just one to two generations considered
in this assessment.

Human populations can deploy some of their inherited physical and
mental abilities in order to adapt to higher tropospheric temperatures,
changed diets or greater crowding. But most of the human adaptations to
changing environment will have to be extrasomatic. Technical fixes can go
far: widespread diffusion of existing techniques and commercialization and
adoption of numerous innovations resulting in higher efficiency and lowered
waste (doing things better) can transform modern farming, energy use,
industrial production and household consumption in profound ways,
making all of them much more compatible with biospheric realities.

But these potentially enormous changes will not succeed without a great
deal of new thinking which will usher in a variety of organizational and
managerial adjustments and transformations, and without extensive individual commitment ranging from a reformulation of life goals to willingness
for co-operative frugality. Doing things as never before and doing without

will have to be the two critical components of the quest for greater compatibility between modern civilization and the biosphere.

DOING THINGS AS NEVER BEFORE

Do all that you know, and try all that you don't . . .
        Lewis Carroll, *The Hunting of the Snark* (1875)

We will not be able to slow down, and eventually to reverse, the world-wide environmental degradation and to avoid unacceptable impacts without a great deal of innovation. This should not evoke images of quasi-miraculous scientific breakthroughs: they will not be the fundamental ingredients of sensible solutions. More than that: excessive pursuit of such strategies and uncritical expectations of their efficacy distract from focusing on the most effective approaches. For example, recent worries about atmospheric warming led to suggestions of such novel technical fixes as dumping of the gas into gargantuan oceanic gigamixers, or incorporating it into ocean phytoplankton whose massive growth would be secured by applications of phosphates and iron, nutrients scarce in surface waters (Marchetti 1977; Roberts 1991). But these are not desirable innovations, merely interesting sci-fi ideas; they are dubious, impractical and economically unacceptable.

What is needed is thoughtful but relatively speedy development of measures which would restrain the growth of affluent consumption and which would enable the poor world's populations to raise the quality of their living to dignified levels with the least achievable degree of environmental degradation. These goals demand that we manage our societies as never before, that we develop a new economics consonant with the long-term maintenance of biospheric integrity. Essential ingredients of this new approach will be better market approaches, more responsive local and national institutions, and a genuine commitment to effective international co-operation. This combination would offer the surest way toward the fastest possible elimination of excessive consumption, toward the fullest possible deployment of most efficient procedures and techniques, and toward the consensual mode of management.

But we must always keep in mind the wide range of scales for effective strategies. Recent attention to the global dimension of environmental degradation must be welcome for its heuristic and mobilizational effects – but its implications for effective management of most environmental problems are dubious. These new approaches are essential for the strengthening of much needed cooperative efforts, but the fashionable preoccupation with the global dimension may also be counterproductive. The adjective remains usually undefined as if its totality would be a perfect guarantee of commonly understood quality. Yet as already noted, all but a few of these problems are world-wide phenomena rather than global changes, and, on reflection, most of them are so time and place-specific, so idiosyncratic, that sensible precepts for effective solutions can come only from careful national, regional and local analyses.

Even in the case of true global effects resulting from climatic change there would be an enormous variation in regional and local effects. The Netherlands, the paragon of a nation threatened by rising seas, would hardly notice a 20 cm rise, nor would it feel threatened by seas 1 m above today's level: its coastal defences are up to 15–18 m higher and its long experience and the strength of its economy would enable it to cope with a gradual rise even higher than one metre. In contrast, repeated massive flooding in Bangladesh caused by cyclonic surges in the Bay of Bengal demonstrates that even without any sea rise the country cannot safeguard a large part of its low-lying territory which truly belongs to the sea. An average sea-level rise of 50 cm would be thus a non-event for the protection of the Dutch IJsselmeer and the Oosterschelde estuary but a major threat for deltas of the Nile and the Ganges (Milliman *et al.* 1989).

Global environmental studies may thus foster a proper focus – or they may obliterate it by trivial or unverifiable generalizations. There is an excellent precedent for this kind of inherent duality: post-1973 energy 'crises' provide an outstanding illustration of some notable benefits but also of the clear futility of shifting the difficult relationship of parts and whole too far in the direction of the whole. After generations of purely particularistic attention to energy matters came an overdue integration of exploration, extraction, conversion, transportation and consumption facets and their integration with environmental and socio-economic consequences. Soon these integrations were carried to international and ultimately to global level. Yet this globalization of concerns has been a counterproductive quest.

Energy supply and conversions are, inevitably, matters of world-wide concern – but not a global problem, just a melange of numerous, and greatly disparate, smaller-scale challenges. After surveying global energy statistics, or modelling the future of global oil prices, what conclusions one can offer to Sichuanese peasants who have too little straw to cook properly before a new harvest and no ties to the world fuel market? Or to rich Saudi families who use water desalinated with the world's cheapest fuel to grow the world's most expensive wheat? Rational management of these matters does not lie in any deeper understanding of global energy resources and markets but rather in adoption of sensible domestic policies going against the ingrained practices and expectations. In the first case it will be the design and diffusion of efficient stoves and promotion of private fuelwood lots, in the other the withdrawal of grotesquely huge subsidies.

Going global is almost always intriguing and sometimes it is undoubtedly very useful. Yet too often those who seem to be acting as guardians of the whole civilization are just privileged providers of abstractions and generalizations which will do nothing to affect local realities. Merely acknowledging local, regional and national differences and peculiarities is not enough: they will have to be at the core of effective control or management designs.

Economics for the biosphere

Only a hopelessly prejudiced analysis of the customary ways in which we run our economies would conclude that the enormous challenges of rational environmental management could be met merely by a continuation of existing practices. Although a number of ways in which we produce and consume in our economies have changed during the past two generations in desirable directions, a growing body of persuasive arguments shows how much more extensive such changes will have to be, and how much further they will have to go (Daly and Cobb 1989; Costanza 1991; Pearce *et al.* 1991). I will introduce here only those innovative approaches which I believe have the best potential for making the greatest difference – but before doing so I must stress the need for an unprecedented willingness to negotiate, to compromise and to co-operate.

The new economics is incompatible with reflexively confrontational approaches. Contending ideas must be encouraged, but litigiousness, so prevalent, especially in North America, and adversarial attitudes must be minimized. Obviously, it is of little use when excellent environmental regulations are disregarded or when promises in international treaties are not backed-up by subsequent investment creating the intended improvements. And it is counterproductive when the only way to bring the desirable change about is through recurrent confrontations and by substituting the judicial rulings for consensual action.

But the fractiousness of modern societies is not an irrefutable proof of lost capacity for co-operation requiring not only effective compromise, but also some real sacrifices. Conflicts and challenges reveal the depth of animosity and intensity of disagreements, but they also demonstrate the potential for meaningful sacrifice and initially unimaginable achievements. Possibilities of determined national commitment entailing major economic sacrifices are naturally best demonstrated by resource mobilization during wars. At the peak of their war-making effort in 1942 the USSR spent almost 70 per cent of its national product on stopping the German aggression, while the British expenditures peaked at 47 per cent in 1943–4 and the American spending reached 54 per cent of the country's GDP (gross domestic product) in 1944 (Harrison 1988).

Making our economies compatible with biospheric realities will not require such enormous sacrifices – but in some ways the task will be more difficult because advances can come only from decades of sustained commitment requiring an unprecedented degree of co-operation. Encouragingly, it is already possible to point out a number of successful co-operative efforts resulting in notable environmental improvements on every scale ranging from local to international. Transformation of Singapore from an overcrowded, insanitary, disease-infested and polluted city of the early 1960s to a model of modern urban development of the 1980s (Jee 1988) required not only capital, management and legislative commitment, but also the active participation of the city state's businesses and citizens.

Lake Erie was so polluted by uncontrolled discharges during the 1950s and 1960s that some scientists considered its degradation irreversible. But after the US–Canadian International Joint Commission made the recommendations for recovery, the two governments co-operated in investing more than US $7.5 billion between 1972 and 1990, mainly to reduce municipal waste discharges and to restore the wall-eye and salmonine fishery. Reduced phosphorus loading improved overall water quality, restored the pelagic ecosystem, and eased the anoxic conditions of the bottom waters (Makarewicz and Bertram 1991). A good multinational example of long-term co-operative commitment requiring willingness to compromise is the European Convention on Long Range Transboundary Air Pollution signed in 1979 after years of negotiations. Its protocol requires the signatories to reduce their total 1980 SO_2 emissions by 30 per cent by the year 1995.

And the Montreal Protocol, signed by more than 30 countries in 1987, and since that time joined by more nations (most notably by China in 1991), is an encouraging example of a speedy international commitment aimed at first reducing and then eliminating the production of CFCs. Its periodical reviews and strengthening of its provision mean that the synthesis of these ozone-depleting compounds will be phased out much faster than initially anticipated. Another notable, although rather controversial, example of effective international co-operation are debt-for-nature swaps in tropical countries (Cody 1988; Page 1988). The combination of immense debts accumulated by many poor countries and worsening terms of trade meant that many debtors found it difficult to pay even interest on their loans. Some banks have tried to avoid a complete loss of their invested asset by unloading debt titles at the best price they could get.

When the Bolivian debt was selling at 15 per cent of its nominal value in 1986, a Washington-based Conservation International spent a US $100,000 grant to buy US $650,000 of debt in order to protect, in co-operation with the government in La Paz, an area of 1.5 Mha (million hectare) of tropical forest in the Beni Biosphere Reserve. Since then a number of debt-for-nature swaps have been completed in Costa Rica, Ecuador and in the Philippines. These arrangements would seem to be win-win combinations. While such swaps may be marginal in terms of easing the enormous indebtedness of poor tropical countries, they may make a great deal of local, national and even global difference. Hypersensitive criticism may see them as infringements of national sovereignty – but they can proceed only after governments and conservationists of debtor countries establish the terms and co-operative arrangements with foreign non-profit organizations managing the investment.

Consensual intentions and co-operative attitudes will be essential in searching for effective solutions. Three basic strategies can be followed by a society intending to make its production and consumption habits consonant with preservation of biospheric services, and using its resources at rates allowing for natural regeneration or innovative substitution: to change the undesirable ways by exhortation, to impose desirable outcomes by regulation, and to transform the whole system by a more realistic pricing of natural resources and services.

Success of the first option would seem to be guaranteed only in a Utopian society. Calls for frugality made by the opinion-making elites would be followed enthusiastically in a well-informed society of altruistic consumers concerned with the welfare of those less fortunate and with intergenerational equity. In a society whose norms are addictive consumption at minimal capital cost, instant gratification on credit, pursuit of personal and special-interest goals to an almost complete exclusion of the public interest, and a growing social malaise, such exhortations sound quite hollow. I cannot think of any better example illustrating the failure of this course than Jimmy Carter's forlorn appeal when he, cardigan-clad, and sitting in front of a flickering fireplace, asked his fellow Americans to conserve energy.

Regulation spans a wide spectrum of approaches. At one extreme are the rigid command-and-control ways of central planning. They are an utterly

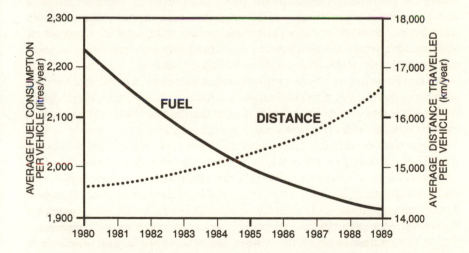

Figure 4.1 A successful technical fix: during the 1980s the fuel consumption of an average American car went down by about 15 per cent in spite of the fact that the average distance driven went up by more than 10 per cent.
Source: Plotted from data in MVMA (1991).

113

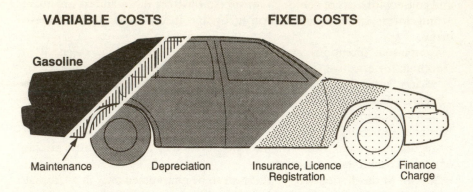

Figure 4.2   A missing incentive: because in 1991 the fuel for an average American car accounted for only one-sixth of the car's lifetime cost, there was hardly any incentive for consumers to buy, and for auto-makers to develop, highly-fuel-efficient vehicles. Breakdown of ownership cost from MVMA (1991) is superimposed on Honda Accord LX, a best seller in its class, and one actually made in USA.

discredited choice: collapse of Communist economies revealed the enormous inefficiencies and horrendous environmental destruction created by economies which had very stringent emission standards and public health norms. Quite opposite are performance criteria imposed after an agreement between legislatures and affected parties. An excellent example has been the gradually tightening regulations prescribing minimum fuel efficiencies and maximum permissible emissions for American cars. They reduced average fuel consumption per vehicle in spite of the fact that average distance travelled per vehicle kept on increasing (Figure 4.1), and they helped to improve air quality in American cities, or, in California's case, they at least prevented an intolerable increase of emissions.

But these regulations could have been much more effective in combination with higher fuel prices. Even in Europe and Japan, where high taxes push the price of gasoline two to three times higher than in the USA, fuel prices are mostly between 30 and 45 per cent of a car's overall lifetime cost; in US in 1991 gasoline accounted for just one-sixth of the total cost of driving (Figure 4.2). This reality largely removes the rationale for buying the most efficient cars available – as well as the incentives for automakers to produce more of such vehicles and to keep improving their performance. An approach combining regulation with incentives should be used when price increases are insufficient to achieve desired environmental effects. A system combining fees for a purchase of an inefficient car and rebates for buying an efficient one should start a desirable trend.

The same principle could be used both among large industrial energy consumers (replacing their motors, boilers, ovens) and in households (for all major appliances). A rewards-only approach may be also effective, providing the companies offering rebates to their consumers can earn profit on this kind of non-traditional investment. For example, Southern California Edison provided more than one million efficient light bulbs free to poor households, offers US $5 rebates to all other consumers buying them in stores (they retail at US $13–18), and it pays its industrial users to buy efficient motors and pumps (Nulty 1991). And the company's commitment to energy conservation will keep on growing because California laws now permit utilities to include investment in electricity-saving capital equipment in their rate base, an innovation which makes it possible to earn profit (at the authorized rate of 13 per cent) from conservation.

Demand-side management programmes including direct installation of more efficient converters, rebates or subsidies for their purchase or installation, and provision of information about opportunities and benefits of efficient techniques should be a major ingredient of an extensive effort which could potentially reduce total electricity demand by between 20 and 40 per cent in a single decade (GAO 1991a). Regulation can also be combined with profit incentives in an innovative approach to cleaning up environmental pollution first suggested by Crocker (1966). Maximum permissible emission rates or average tolerable concentrations should be legislated on the basis of the best scientific evidence, but the compliance should be achieved by market-based incentives through emission permit trading (Oates 1988).

This approach is based on a common fact that pollution control costs will vary among companies both within and between affected industries. Instead of a centrally-dictated proportional allotment of control quota based on legislated targets, this arrangement allows the market to set the final allocation. Of course, a regulatory body still sets the overall control target – but then it issues tradable (auctionable, too) emission permits to affected companies. Given the differences in marginal control costs companies with relatively low abatement costs could find it profitable to reduce their emissions well below the level assigned by proportional allotment – and sell (or auction off) their excess emission permits to polluters facing higher marginal control costs. Consequently, the overall cost of mandated emission controls should be lower than with a rigid legislative distribution.

This approach has broad, and often highly rewarding, applications. A utility generating electricity in a number of large thermal stations can equip its latest coal-fired plant with the best available flue gas desulphurization units – and transfer the plant's SO_2 emission allowances to an old station which will operate during its few remaining years without any controls. And emission permits can be traded cost-effectively, not only among different companies of the same industry (for example, refineries in New York-New Jersey area), but also different industries (SO_2 from power plants and smelters) and different countries (in a European effort to reduce acid deposition, possibly in a global quest for lower CO_2 emissions).

Encouragingly, tradable pollution allowances are to be the key to reducing

US SO_2 emissions by 10 Mt (million tonnes) by the year 2000 as mandated by the 1990 Clean Air Act. But as useful as innovative regulations can be, they are only the second best to a more efficient functioning of a market based on a more realistic valuation of natural resources and services: paying for the real cost of environmental degradation provides the best hope for the success of the new economics.

Paying the real price

The challenge of bringing market prices in line with hidden burdens of environmental degradation is so extraordinarily complex that it will be with us for generations. Although the adoption of a new set of economic ground rules must be done gradually, we must accelerate this difficult process by extending it to a wider range of commodities and services, and by trying to come up with progressively fuller accounts. In practice this superior approach will need important social support: markets float on a sea of regulation, and they cannot function well in semi-literate and misinformed societies.

Perhaps the best way to start moderating the excesses of affluent societies is through a gradual introduction of more realistic energy prices. The whole world lives with unrealistically low energy prices, but because in rich societies fossil fuels and nuclear and hydro electricity energize virtually all economic activity, the greatest declines in energy use would come from internalizing first the principal externalities of energy use in Europe and North America: direct financial subsidies, economic dislocations, environmental degradation, health costs, and military expenditures.

Because energy industries and consumption of fuels and electricity have enormous local, regional and national economic effect, western governments have been forthcoming with considerable research funds and with generous tax credits aimed at advancing, stimulating and subsidizing development and production activities. Costly examples range from decades of funding for nuclear fission research in the USA, UK and France to Canadian taxpayer's money poured into dry Arctic wells drilled in the quest for future energy self-sufficiency. Preferential treatment of some energies – be it by subsidies to specified commodities or processes or by favouring resource development in particular locations – can, by shifting investment and labour opportunities, create sizeable economic dislocations. These changes may lead to net economic gains – but also to substantial losses once such preferences, usually granted by a legislative fiat, are removed.

As already noted, environmental impacts of energy production and use embrace an enormous range of undesirable changes. Spectacular accidents – such as the destruction of Chernobyl's unshielded reactor in 1986 or the crude oil spill from *Exxon Valdez* in 1989 – capture public attention with images of horrifying damage, but the effects of cumulative long-term

116

changes are generally far more worrying. Those which have already caused serious local or regional degradation include photochemical smog and acid deposition, and their costs would be dwarfed by potentially destabilizing global climatic change induced by emissions of greenhouse gases. Fossil fuel combustion is by far the largest source of CO_2, and production of natural gas and emissions from coal seams are among the largest sources of anthropogenic CH_4.

Occupational exposures cause a number of life-shortening diseases among workers in energy industries (miners' black lungs are perhaps the best known case), but the most serious public health risks come from long-term, low-level breathing of air pollutants from fossil fuel combustion (particulates, SO_2, NO_x, unburned hydrocarbons, lead) or formed by subsequent reactions in the troposphere (sulphates, nitrates, ozone). Risks of very low exposures to ionizing radiation from fission power-plants and long-term consequences of higher doses released by accidents are the second-largest class of health externalities tied to modern energy uses. While health risks and environmental changes are universal, economic dislocations and subsidies matter only among major fuel and electricity producers, and the USA alone has the dubious distinction of devoting enormous military expenditures to assuring a free flow of crude oil from the Middle East.

During the past two generations western economies have already made some notable advances toward more realistic energy pricing. Fifty years ago most underground coal-miners were exposed to unlimited levels of dust and did not have any disability insurance or compensation benefits in their contracts. Most power-plants burning that coal did not have any air pollution controls, and there was hardly any concern about pollutants' effects on visibility, ecosystems and public health. Today governmental regulations prescribe tolerable dust concentrations in mines and disability and compensation provisions in union contracts help to blunt the blow of occupational sickness.

Effective pollution controls remove virtually all fly ash and more stations are equipped with flue gas desulphurization. Large utilities, acting alone or in consortia, conduct and support health and environmental research and are investing in alternative methods of securing supply, either by promoting non-fossil generation or by community conservation campaigns. Well-publicized examples of internalizing costs in oil industries involve large tankers. Steep increases in insurance premiums for giant tankers reflecting the risks of major crude oil spills have been a key factor in stopping a further growth of these huge vessels. And after the grounding of *Exxon Valdez* the company spent directly about US $2 billion on the clean-up and paid half as much to the state of Alaska. Costs of restoring waters and (at least superficially) rocky shores of the Prince William Sound were thus internalized to a degree unprecedented in oil spill accident.

In the future a number of uncertainties and complexities will make it exceedingly difficult to internalize precisely those hidden burdens of energy use which appear to have the highest external cost – increased morbidity and excessive and premature mortality resulting from chronic low-level exposures to air pollutants, and cumulative ecosystemic changes cause by human interference in grand biospheric cycles. Naturally, the quest for realistic pricing should not stop with energy. Prices of wood sold for timber or pulp do not express the real cost of clearcutting: higher rates of soil erosion, loss of water storage capacity, nutrients and organic matter, increased downstream siltation, the impossibility to re-establish the forest in many vulnerable locations, and decline in biodiversity. Analogically, prices of water used for irrigation do not reflect the long-term costs of aquifer depletion, soil salinization, or degradation of downstream aquatic ecosystems.

Water pricing in arid regions with recurrent water scarcities actually offers some of the worst examples of current underpricing because it does not cover even the costs of delivery. In California farmers using water delivered by the Central Valley Project repay their share of building costs over a period of 50 years with interest-free loans, and most of them have been getting it at just about 10 per cent of the actual supply cost (Gottlieb 1991). During the late 1980s, a decade of extensive drought and chronic urban water shortages, the state-fixed price for water in China's capital was merely a quarter of its delivered cost, and the typical cost of China's irrigation water was mostly between 5 and 20 per cent of the actual cost (Smil 1993).

A more realistic pricing of natural resources and services should be complemented by new sets of national economic accounts which would go beyond the traditional valuation restricted to production of goods and services and would take into consideration the cumulative changes in natural assets. These valuations would expose the fallacy of prevailing guidelines for the standard system of national accounts which ignore the deterioration of the environment and which treat the depletion of natural assets as current income boosting GDP totals. No society can keep running down its irreplaceable natural capital, treat this process as income, and hope to survive for long.

While standard accounts may show high rates of economic growth, more comprehensive assessment may reveal a stagnation or even decline. A growing number of economists has been working on the preparation of adjusted national accounts which treat the environment, at least partially, as capital. The most comprehensive exercise for the USA has been Daly and Cobb's (1989) per capita index of sustainable economic welfare (ISEW) which includes costs of commuting, car accidents, urbanization, water, air and noise pollution, loss of farmland and wetlands, depletion of non-renewable resources and an estimate of long-term environmental damages. Since 1950 the American per capita GNP (gross national product) has always resumed its growth, while the ISEW stopped growing after the late 1960s, and it has been declining since the mid-1970s: while per capita GNP

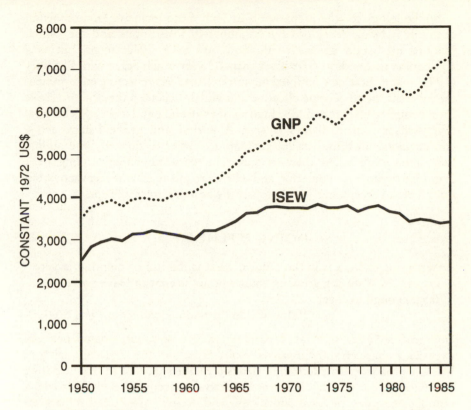

Figure 4.3 While American GNP doubled between the early 1960s and the mid-1980s, the index of sustainable economic welfare calculated by Daly and Cobb (1989) has hardly changed. This is just one possible approach to calculating alternative indicators of real wealth which differ substantially from standard GNP figures.
Source: Daly and Cobb (1989).

rose by just over 20 per cent between 1976 and 1986, per capita ISEW fell by 10 per cent (Figure 4.3).

If Daly and Cobb's (1989) estimates are basically correct, then we have to conclude that the economic growth of the past generation has been a delusion resulting from irrational accounting procedures which simply ignore exhaustion and degradation of natural capital. Less drastically, in a re-evaluation of national accounts for Costa Rica, the World Resources Institute found that more than a quarter of the country's pre-1989 economic growth disappears when adjustments are made for the depreciation of forests, soils and fish stocks (Repetto *et al.* 1991). This reduction would be even greater if the new accounts would have included the value of lost forests as wildlife habitats, tourist attractions (Costa Rica has a thriving eco-tourism business), water storages and repositories of unique biodiversity.

119

Methodology for the preparation of such more inclusive environmental accounts is being developed by a number of institutions and the United Nations may soon use such valuations not as a replacement, but as a companion to standard GDP assessments. I very much agree with El Serafy (1991) that if these new national income calculations reflecting environmental concerns are to become effective and widely accepted they cannot strive for a complete inventory and valuation of environmental assets. Such a goal will perhaps remain always elusive. A partial and gradual approach – concentrating on those variables which are easier to quantify and adding new items only when there is sufficient scientific understanding of a degradative process in question and when a requisite accounting innovation finds a broad acceptance – is obviously the best practical solution.

DOING WITHOUT

When the cholera is in the potato, what is the use of planting larger crops? . . . Want is a growing giant whom the coat of Have was never large enough to cover.

Ralph Waldo Emerson, *Conduct of Life* (1860)

Unprecedented access to an expanding variety of foods, possessions and experiences has certainly improved many aspects of the physical quality of life, and broadened intellectual horizons for large shares of the participating populations. But this development has also damaged and even destroyed a number of other physical amenities, and it has substituted superficial knowledge for real understanding. Excessive consumption of previously scarce foods (meat, fats, sugar) has been an important factor in the rise of many diseases. Production of this nutritional excess and saturation of markets with a numbing variety of goods and services has lowered the quality of life, first for people immediately affected by rising extraction and conversion of energy and materials, and recently for everybody through worrying possibilities of global environmental changes. And the surfeit of possessions and pastimes has brought more, not less, social disintegration, ranging from the abuse of children and women to growing functional illiteracy.

Clearly, a strong case can be made for protecting the environment, not only by using natural resources more efficiently, but by simply consuming less energy, amassing fewer possessions and opting for more benign pastimes. Profound cultural and historical differences, disparities in resource endowment, and entrenched infrastructural peculiarities, preclude any normative generalizations, but closer looks at the links between important consumption indicators and the quality of life indicate that all affluent societies, or roughly a fifth of mankind, have a surprisingly large potential for doing without.

120

Consumption and the quality of life

Higher consumption is commonly seen as the foundation for improving the quality of life – but international comparisons reveal neither any strong correlations nor any pre-ordained paths of development. The most obvious reason for this looseness is the inherently multifactorial nature of the quality of life. The concept ranges from narrowly conceived personal well-being to wider environmental and social setting (including a large array of natural and man-made risks ranging from air pollution to crime) – and it also embraces the vast mental aspect of human development (starting with basic education and including the exercise of political and religious freedoms).

In spite of their numerous weaknesses, per capita GDP values do convey a fundamental message of economic achievement – but neither a solitary figure nor a multivariable index can be adequate stand-ins for an assessment where many hard-to-measure qualities and perceptions are more important

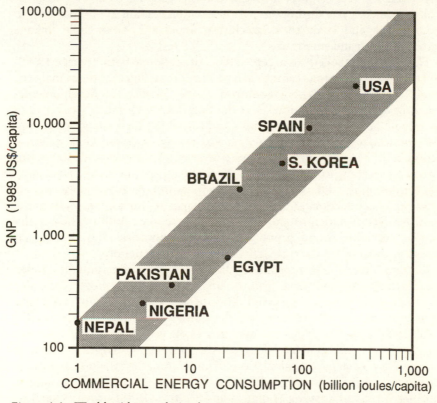

Figure 4.4 World-wide correlation between GNP and primary energy consumption is above 90 per cent. This has led to a persistent misunderstanding that higher income levels are inextricably bound with correspondingly rising primary energy consumption.

Source: Plotted from 1989 data in World Resources Institute (1992).

than readily available quantitative information. In fact, many studies have shown little connection between subjective appraisals of the quality of life and personal satisfaction and the objective socio-economic indicators (Nader and Beckerman 1978; Andrews 1986).

No link between resource consumption and the standard of living is more obvious than the one between national affluence and the flow of primary energy. In order to generate an average per capita GNP of over US $20,000 during the early 1990s, the United States consumed annually about 300 GJ (gigajoules) of primary energy for each of its inhabitants. In contrast, the precarious maintenance of the US $150/capita Bangladeshi economy drew on inputs of mere 2 GJ of primary commercial energy per capita. There seems to be a nearly perfect correspondence between these extremes as the two orders of magnitude higher energy inputs generate two orders of magnitude higher GDPs – and the relationship seems to hold well at intermediate levels (Figure 4.4). Correlations ranging from 0.9–0.95 for world-wide data sets during the past generation explain 80–90 per cent of the variance and leads to a conclusion about the inextricable linkage between wealth and energy use.

This link was labelled 'an accepted tenet of economic faith' (Jensen 1970), and the high correlation seems to offer a fine forecasting tool to estimate the future energy needs of poor countries as they retrace the developmental stages covered by rich nations. But the link is actually much weaker than commonly believed: data limitations and mistaking high energy consumption for a high quality of life explain this reality. Standard GDP figures – expressed in national currencies and converted to a common denominator by using prevailing exchange rates – have several specific weaknesses (even when ignoring the dubious nature of the measure). Available data exclude subsistence production and barter and they also miss often substantial contributions of underground economic transactions. Both of these activities are generally more prominent in poor countries, and hence their omission undervalues the real GDP of industrializing nations.

Exchange rates reflect primarily the prices of internationally traded commodities and may have little or no relationship to those parts of a country's economy not involved in foreign exchange. And rapid and substantial currency fluctuations which can boost or lower a country's GDP in US $ easily by 10–20 per cent within a single year have been a recurrent problem ever since the free float started in 1972. Corrections through purchasing power parity calculations remove some distortions, mainly by effectively raising the adjusted GDPs of poorer countries. The longest (1950–85) retrospective by Summers and Heston (1988) provides such an adjustment based on average relative prices for 130 countries. But these adjustments cannot avoid arguable constructs: it is impossible to specify an average global food basket, typical expectation of adequate housing or a satisfactory make-up and frequency of leisure expenditures.

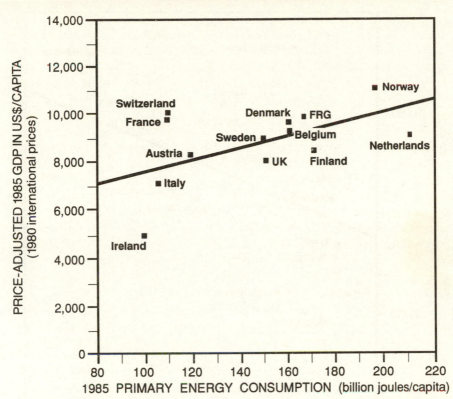

Figure 4.5 Correlation between GNP and primary energy consumption is rather weak for a relatively homogeneous group of industrialized European countries even when their economic products are adjusted for purchasing power parities.
Sources: Energy values from UN (1985); PPP-adjusted GDPs from Summers and Heston (1988).

Very high correlations for sets spanning values ranging over two orders of magnitude weaken substantially when performing separate regressions for more homogeneous groups of economies. Huge differences of energy inputs, and clear lack of any meaningful energy-GNP correlation, are obvious when looking at European data (Figure 4.5), and similarly diffuse patterns can be found when analysing middle-income and low-income economies (Figure 4.6). Another important fact is that the link between GDP and energy keeps changing, that energy intensities of individual economies (amount of energy used per unit of GDP) tend to rise in the early stages of their industrialization and decline afterwards. This fall has been especially pronounced for countries whose primary energy consumption was originally dominated by coal: UK and USA have experienced a clear long-term decline of energy/GDP intensity (Humphrey and Stanislaw 1979; Smil 1991a).

Energy intensities of economies without major coal production, such as

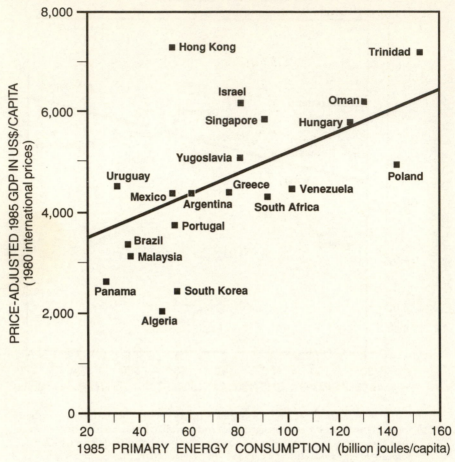

Figure 4.6 Another illustration of a weak GNP-energy correlation is shown for the World Bank's group of middle-income economies.

Sources: Energy values from UN (1985); PPP-adjusted GDPs from Summers and Heston (1988).

Japan or Sweden, have never risen so high as in major coal-producing nations – but these countries have also experienced long-term declines of this critical ratio (Figure 4.7). And this trend has been most recently retraced by impressive reductions in modernizing Chinese economy (Figure 4.8). Consequently, future energy demand of the poor world's growing population aspiring for a better quality of life will have to be substantially above today's low rates – but it is certain that it will not replicate the secular rise and fall of energy intensities of the old industrial powers. Appropriate levels of energy use can be derived by narrowing the search to fundamental correlations.

Any sensible definition of the good life must contain such more-or-less

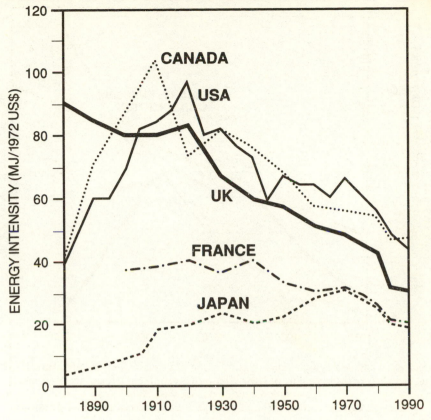

Figure 4.7 Long-term decline of energy intensity in major western economies displays expected national idiosyncracies, as well as the general downward trend after the increases during the early stages of industrialization.
Source: Based on Smil (1991a), BP (1991) and World Bank (1991).

adequately measurable variables as health care, nutrition, housing and education. As long as one looks across the whole developmental spectrum there is little doubt about the existence of fundamental, robust relationships between the levels of per capita energy utilization and the *physical* quality of life characterized by these variables: the two orders of magnitude separating the total per capita energy use of an average city dweller of a rich western nation from that of a typical Bangladeshi villager are clearly the key explanation of the vast material and intellectual gap between the two societies.

On the other hand, few among the unquantifiable or intangible ingredients of a good life require high energy inputs. With the exception of such high-power, quintessentially American, pastimes as power boating, aimless car cruising and stock-car, dragster or dune buggy racing, common leisure

125

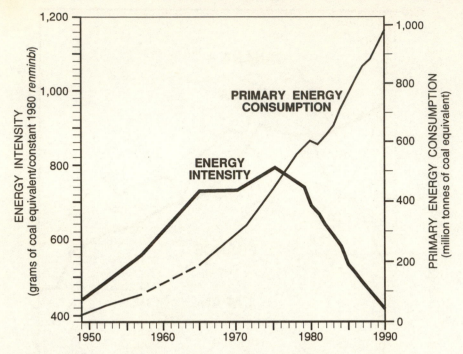

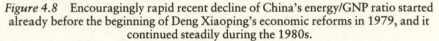

Figure 4.8   Encouragingly rapid recent decline of China's energy/GNP ratio started already before the beginning of Deng Xiaoping's economic reforms in 1979, and it continued steadily during the 1980s.
Source: Calculated from data in SSB (1991) and from information in Smil (1988).

activities demand either no additional metabolism or only a modest increase in food consumption, and only slight inputs of embodied energy in producing books, recordings, table games, gardening and sports equipment and long-lasting electronic gadgets. Rich legacies of physical contests, games, hobbies, literature and music predating the huge increase of energy consumption in modern societies testify to a lack of any strong and consistent link between rich and sophisticated pastimes and energy use.

And it is also indisputable that there has been no obvious link between high energy use and political freedoms: historical and cultural peculiarities have been much more important factors. Consequently, the United States adopted its visionary constitution while the society was a relatively low consumer of wood – while Germany slid into aggressive militarism and later into Fascism just as it became Europe's leading consumer of coal. And the one-party dictatorships of the former Soviet empire used significantly more energy than the West European democracies. Consequently, health, nutritional and environmental variables characterizing the physical quality of life should reveal more about the links between energy use and higher standard of living.

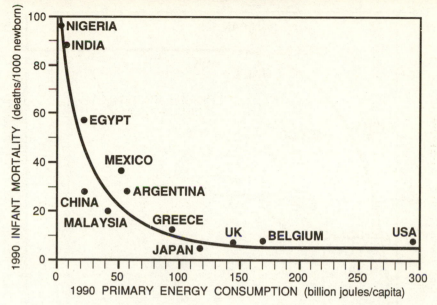

Figure 4.9 Infant mortality drops precipitously with increasing energy consumption between a few and 50 GJ/capita a year, but the gains are only modest with energy use rising to between 50 and 100 GJ/capita, and none at all above that rate.
Source: Plotted from data in World Resources Institute (1992).

An effective way to find minimum energy needs compatible with a good life is to plot key quality-of-life indicators against average per capita energy use: the most remarkable result of this exercise is the existence of unmistakable inflection points at between 40–50 GJ/capita: increasing per capita energy use beyond this range brings, first, greatly-diminishing returns, and then a levelling-off with no gains at all. This is true about several measures of physical quality of life, most notably about infant mortality (Figure 4.9) and life expectancy (Figure 4.10). Indicators of intellectual development are much harder to interpret. But when literacy rates in excess of 80 per cent are taken merely as indicators of general accessibility to basic education, the corresponding annual energy consumption is, once again, between 40–50 GJ/capita.

If existential and educational basics are covered with per capita energy uses of 40–50 GJ/year, only marginal increases may be needed to approach the world's best levels. Life expectancy of about 75 years and infant mortality below 10 require energy use of around 70 GJ/capita, as do literacies above 90 per cent. Perhaps the most revealing measure of modern educational opportunities is the share of young adults receiving post-secondary education – and all but a few countries enrolling more than 20 per

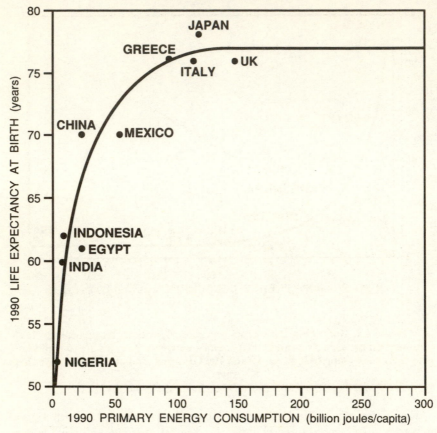

Figure 4.10 Life expectancy at birth rises from just over 50 years to about 70 years as the average annual per capita energy use goes from a few GJ to about 50 GJ, but the gain is only between 4 and 7 years with the rise to around 100 GJ, with no gains afterwards.
Source: Data source as in Figure 4.9.

cent of potential students consume annually at least 70 GJ of primary energy per capita (Figure 4.11).

All primary energy consumption rates arise from conversion efficiencies prevailing during the late 1980s and the early 1990s. A 25 per cent efficiency gain over a generation would lower the annual rate of primary energy needs assuring high levels of post-secondary education to about 50 GJ/capita. As this was the global average at the beginning of the 1990s, egalitarian division of the global primary energy production combined with a substantial but realistic increase of average conversion efficiencies could supply everybody with fuel and electricity guaranteeing a rather high standard of living.

After all, in terms of effective energy services per capita, primary energy

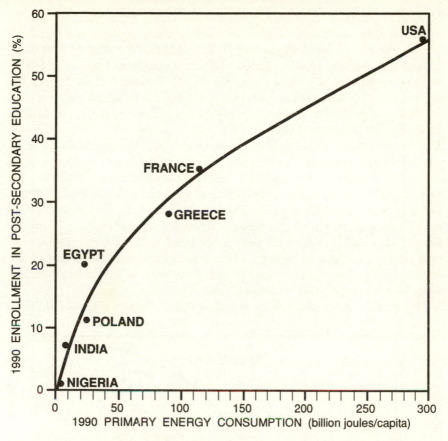

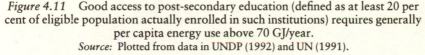

Figure 4.11 Good access to post-secondary education (defined as at least 20 per cent of eligible population actually enrolled in such institutions) requires generally per capita energy use above 70 GJ/year.
Source: Plotted from data in UNDP (1992) and UN (1991).

supply of 50 GJ/year in 2010–2015 would be an equivalent of at least 80 GJ/year during the first half of the 1960s – or basically the level achieved at that time by France (average French per capita primary energy consumption went from just over 70 GJ/year in 1960 to about 90 GJ/year in 1965). Who could object to sharing the quality of life enjoyed by the French in the early 1960s? Definitely not Haitians, Egyptians and Bangladeshis – but almost certainly the French themselves, and together with them the inhabitants of all other affluent societies.

Lifestyles, expectations and global consumption

This resistance would spring from perhaps the most remarkable fact about the modern western mass affluence: from its surprisingly recent origins.

129

Naturally, different countries have moved along their peculiar developmental paths, and all indicators of comfortable living have not marched in step, but it is a valid generalization to conclude that the mass middle-class affluence dates only since the 1950s in North America and since the 1960s in Europe, with Japan joining during the 1970s. To demonstrate this reality it is better to avoid American or Japanese examples: the first instance is one of clearly unparalleled average per capita consumption, the other case is obviously one of *nouveau riche*.

Using France, a nation of old culture and no small technical prowess, with a generally acknowledged high quality of life, is a much more representative choice. In retrospect, the country's recent state of relative poverty seems surprising even to such a knowledgeable observer as Antoine Prost (1991). In his sweeping analysis of public and private spheres in modern France he noted how the 1954 census revealed a striking image of the primitivity of French housing – less than 60 per cent of households had running water, only one-quarter had an indoor toilet and only one in ten had a bathroom and central heating – and concluded that 'it is hard to believe that we are describing a situation that existed only thirty-five years ago'.

The next 20 years saw a huge transformation. French statistics (INSEE

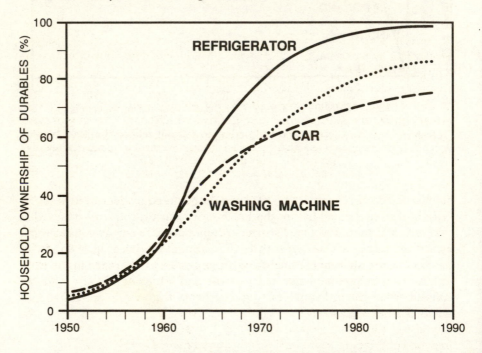

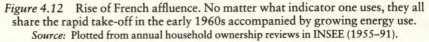

Figure 4.12 Rise of French affluence. No matter what indicator one uses, they all share the rapid take-off in the early 1960s accompanied by growing energy use.
Source: Plotted from annual household ownership reviews in INSEE (1955–91).

130

1955–91) show that nearly all attributes of household affluence became widely available during the 1960s and the early 1970s. At the beginning of the 1960s only one-fifth of all households had central heating, less than a third had a bathroom, refrigerator or washing machine, and only 40 per cent had a flushing toilet inside. By the mid-1970s refrigerators were in almost 90 per cent of households, toilets in three-quarters, 70 per cent had bathrooms and about 60 per cent enjoyed central heating and washing machines. Since then all of these possessions have become virtually universal, and three-quarters of all families also own a car, compared to less than a third in 1960 (Figure 4.12).

Inevitably, this growing material affluence had to be reflected in rising demand for energy: between 1950 and 1960 the total per capita consumption of primary energy rose by almost exactly 25 per cent, but between 1960 and 1974 it soared by over 80 per cent (UN 1976). And while between 1950 and 1990 the per capita supply of all fuels more than doubled, gasoline consumption rose nearly sixfold and electricity use went up more than eightfold. These trends are representative of the three general energy consumption shifts associated with the growing western affluence: rapid post-1960 rise of total energy use, a strong demand for liquid fuels needed to power greater personal mobility, and an even stronger demand for electricity.

The only way to energize a more secure global civilization without massive production increases could be then achieved only with deep consumption cuts – anywhere between 30 and 80 per cent of the early 1990s rates – in rich countries. Such reductions may seem quite unrealistic – and yet they would not be impossible. Reducing energy use would seem to run against the evolutionary quest for maximization of energy flows which has marked the growing complexity of the biosphere and the ascent of fossil-fuelled civilization. But there are no natural obstacles to such a reversal. Energy flows are absolutely essential for the survival of organisms, but their modification and differential utilization is determined by properties intrinsic to the organisms. Energy flows are not the determining factors of biospheric organization and they cannot explain the existence of organisms and their variability.

I endorse Brooks and Wiley's (1986) suggestion to regard living organisms as 'physical systems with genetically and epigenetically determined individual characteristics, which utilize energy that is flowing through the environment in a relatively stochastic manner'. This reality has profound consequences for assessing the prospects of industrial civilization. Its unprecedented dependence on huge energy flows can be seen as a result of organismic and social evolution sequestering steadily larger flows of energy in conformity with Lotka's (1925) law of maximum energy:

> In every instance considered, natural selection will so operate as to increase the total mass of the organic system, to increase the rate of

131

circulation of matter through the system, and to increase the total energy flux through the system so long as there is present and unutilized residue of matter and available energy.

But if stochastic use describes better the evolutionary reality then the growing energy use cannot be equated with effective adaptations and we should be able to stop and even to reverse that trend. This should be easier given the clear indications that maximizing of power output has become existentially counterproductive. Indeed, higher energy use by itself does not guarantee anything except greater environmental burdens (Smil 1991a). Opportunities for this grand transition to less energy-intensive society can be found primarily among the world's pre-eminent abusers of energy and materials in Western Europe, North America and Japan – and some could be surprisingly easy to realize.

These countries have reached near-saturation with heavy-duty appliances whose efficiencies can be greatly improved by innovative design and electronic controls, while new electronic devices have generally low power needs: consequently, it should be possible to make deep long-term cuts in electricity demand in all rich western nations. Former Communist economies could also reduce electricity generation: existing supply shortages have been caused by inordinately wasteful consumption in the disproportionately large heavy industrial sector, a reality which is now being remedied by industrial restructuring.

In contrast, production, ownership and use of passenger cars show no signs of saturation, merely some cyclical changes. The world-wide production of cars rose from about 28 million in 1980 to 35 million by 1989; by the late 1970s less than 30 per cent of US households had two and less than 10 per cent owned three cars, while a decade later the two shares were, respectively, close to 40 and over 15; and the average distance travelled annually by an American vehicle went up by nearly 15 per cent during the 1980s, a trend reflecting continuing residential sprawl and more recreational driving (MVMA 1991). And the still-far-from-saturated markets in the former Communist countries mean that the growth of their car ownership and use will be fairly strong.

Breaking the universal infatuation with cars will not be easy. As Boulding (1974) observed so acutely, the automobile is especially addictive: 'It turns its driver into a knight with a mobility of the aristocrat and perhaps some of his other vices'. But there is nothing inherently beneficial about that addiction, there is no fundamental link between a high quality of life and car ownership and extensive use. In fact, enormously costly externalities associated with mass driving – ranging from rising totals of casualties and injuries to severe air pollution and excessive noise – should justify a resolute effort to move away from the automobilization of modern societies.

Doing without does not have to mean deprivation and sharp declines in the standard of living; rather it means a separation of obviously unsustainable material expectations from both the personal and societal perception of a high quality of life. And while it would be naïve to think that this transition to sanity could be achieved without any individual, community or national sacrifices, it is also necessary to appreciate formidable possibilities for higher efficiencies, reduced waste and smarter management.

DOING THINGS BETTER

> They therefore do not only use much less woollen cloth than is used in other countries, but also the same stands them in much less cost.
> Thomas More, *Utopia* (1516)

Frugality and rational consumption were undoubtedly easier to accept and to perpetuate in Moore's imaginary realm – but the advantages of conservation and efficient production should actually hold even more appeal for modern civilization. Two reasons explain the difference. Where the Utopians had to do with one colour (natural) of their woollen cloth, our inventiveness gives us usually a much greater choice of techniques, products and approaches within the confines of rational management. And whatever their savings were, we can surely surpass them as many gains we can realize with already fully commercial techniques represent radical departures rather than merely marginal advantages.

I will concentrate once again on existential necessities: growing of food, conversions of energy and consumption of water and materials. Doing these things better means farming without excess, wasting less water, improving the efficiencies of fuel extraction and combustion, and reducing the use of wood, metals, and synthetic materials. These strategies – relying on prevention, anticipation, and reduced consumption – are generally superior to the reliance on clean-up and control techniques grafted on dominant inefficient processes and arrangements.

Out of the multitude of possible improvements I will select those technical and managerial advances which hold the highest promise for making the world-wide difference. This choice is fairly easy for innovations that have already entered the commercial phase or whose diffusion is imminent – but restraint is needed when picking the potentially most rewarding emerging improvements. The challenge is to separate a number of realistic prospects from appealing (and hence often highly publicized) but much less likely (or even highly improbable) changes.

Sometimes that task is easy: energy scenarios for the coming generation contain both photovoltaics and solar satellite power, but the probability of extensive applications of photovoltaics for commercial electricity generation is incomparably higher than the construction of geostationary satellites broadcasting the energy to the Earth as microwaves. But more often we simply do not know: after a generation of intensifying research we are no closer to achieving a working symbiosis of nitrogen-fixing *Rhizobium* bacteria with cereal crops – but how can one categorically exclude a breakthrough within the next generation? Yet these uncertainties do not matter that much as far as the next 10–25 years are concerned: merely using already existing abilities during that period would bring substantial reduction of environmental degradation throughout the rich world, and notable slowdown of undesirable changes in industrializing countries.

Farming without excess

A simple fact first: the prevailing way of cultivation is clearly unsustainable on the civilizational scale; it will not last thousands of years as did a variety of traditional practices. But unsound modern agronomic practices do not have a viable alternative in the return to traditional farming characterized by complex crop rotations, heavy organic recycling and reliance on animate labour. This is not because organic agriculture could not support high yields. Stanhills's (1990a) extensive review of the comparative productivity of organic and conventional farming showed that the yields from the former practice were, on the average, no more than 10 per cent less than from the latter. Moreover, in almost one-third of cases they were actually higher.

Two principal reasons make organic farming an impossible universal remedy. Even if the citizens of industrialized nations would reduce their extraordinarily high meat intakes (and hence release large areas of farmland, now used for growing animal feed, from the cultivation), cropping the remainder in traditional ways would be possible only with an appreciable de-urbanization of western societies, with a substantial increase of field and farmyard labour and with a large share of farmland devoted to growing feed crops for draft animals. But however impractical and unlikely, such a shift could be accomplished without imperilling the adequacy of typical western nutrition.

In contrast, in all land-scarce poor countries where intensive cropping produces largely vegetarian, and for too many people still too often deficient, diets, a return to traditional farming would be possible only if it would be accompanied by drastic declines in the current populations. In the China of the early 1990s, inorganic fertilizers provide about 60 per cent of all nitrogen, and even if the Chinese would reduce their largely vegetarian protein intakes to existential minima, synthetic nitrogen would still be needed to grow two-fifths of the country's food harvest (Smil 1992b).

Consequently, the only way out is to evolve a more efficient and ecosystemically more acceptable set of cropping practices. This new agronomy has been emerging since the 1970s with a relatively rapid expansion of a wide variety of reduced tillage practices (Gebhardt *et al.* 1985), with the diffusion of efficient irrigation techniques (O'Mara 1988), and with attempts to lower the applications of pesticides, to reduce the extent of monocropping, and to expand organic recycling (USDA 1980; Bezdicek *et al.* 1984). These changes have not been simply by-products of recently fashionable environmental concerns: they arise from a growing realization of the unsustainability of many widespread post-1945 practices (NRC 1989; Edwards *et al.* 1990).

Transition to a truly sustainable farming will have to be gradual, but there are no fundamental obstacles to the immediate introduction of many better approaches: to farm without excess is the necessary first step. No national agriculture conforms today to this desirable pattern. Monocultures are a norm on every continent; cultivation of leguminous food crops in complex rotations with staple grains has been steadily declining everywhere with the exception of India; with planting of green manures and recycling of crop residues and animal wastes in decline, organic matter has become a marginal source of nutrients in comparison with massive applications of synthetic fertilizers; irrigation is often either excessive or highly inefficient; and cultivation of land which should have remained under grasses or trees is a growing cause of soil erosion and desertification.

Although the multitude of factors will dictate many specific approaches, there is good agreement about the universally applicable contents of successful agro-ecosystemic management (Pimentel *et al.* 1989; Munson and Runge 1990). They include widespread adoption of crop rotations, with special emphasis on leguminous species; extensive recycling of organic wastes maintaining soil quality; more efficient applications of synthetic fertilizers; maximum practicable use of reduced tillage, or even no-till, methods; and greatly reduced waste of irrigation water (a concern treated in a separate section later in this chapter).

Many no-cost or low-cost decisions are only a matter of appropriate know-how. Merely knowing when to plant can be very valuable. Seeding beyond a narrow optimum period may cut American corn yields by 125 kg/ha per day, and in Kenya's Eastern and Central Provinces the difference may be as much as 170 kg/ha a day (Haugerud and Collinson 1991). Where the choice of cultivars is readily available, the optimum selection of high-yielding varieties resistant to disease or insects can make a substantial difference with the same rate of nutrient and pesticide inputs. Yield differences in US can be around 3 t/ha for corn and about 1.3 t/ha for soy beans. Knowing how to plant also matters: optimum density of corn seeding raises yields by as much as 2.5 t/ha, while narrowing soy bean rows can bring another tonne per hectare.

Declining diversity of cropping can be seen equally well in land-rich and still largely extensive agricultures and in land-poor intensively cultivated agro-ecosystems. In 1949 Chinese peasants planted wheat and rice on 43 per cent of all grain-growing area, in 1990 on 55 per cent. Sorghum, millet, barley and oats were planted on more than 29 Mha in 1952 (a fifth of all sown land), but by 1985 their area had shrunk to less than 10 Mha. Cultivation of such traditional Chinese species as millet, mung beans and red beans has become very rare. The planting of soy beans, China's leading leguminous crop, increased after the privatization of farming in the early 1980s, but the total area is still one-third below the peak reached in 1957, and even below the record pre-Second World War extent, so that the 1990 annual per capita availability of soy beans is only two-thirds of that a generation ago.

These world-wide trends toward monocultures and away from cultivation of legumes and green manures will have to be reversed. Rational cropping must include rotations: they were the mainstay of all traditional agro-ecosystems and their rapid decline, and sometimes outright abandonment, during the past two generations should be seen as an undesirable aberration. The environmental advantages of rotations have been well documented by long-term field experiments (Higgs *et al.* 1990). They include, above all, reduced soil erosion and water runoff, nutritional benefits of symbiotic nitrogen production by legumes, better soil tilth (largely because of the forage crops used in rotations), and the interruption of weed, insect and crop disease cycles.

Successful rotations must include leguminous plants and so there should be much more attention given to the productivity of edible varieties which also make substantial contributions to local protein supply (Matthews 1989). An outstanding recent example of this approach is the release of an improved strain of mung bean by the Asian Vegetable Research and Development Centre. Between 1985 and 1990 this cultivar was planted on over 360,000 ha in China, outyielding local varieties by about 50 per cent and bringing a major economic benefit (Harris 1991).

Applying just the right amount of fertilizers brings both higher profits and environmental benefits. Losses can be cut in both rich and poor countries by a combination of soil testing, more precise applications and good agronomic practices. The most effective measures concern the timing of application and the placement of fertilizers: synchronizing nitrogen applications with periods of the greatest nutrient need and incorporating the fertilizer into the soil are the two key improvements (Munson and Runge 1990). Improvements in fertilizer efficiency could already be seen in the US record of the 1980s. While the total applications of phosphorus peaked in 1977, and those of nitrogen and potassium in 1981, average yields of principal field crops continue to increase.

Maintaining sufficient organic matter content in intensively farmed soils can be achieved by a combination of crop rotations, regular manure applications, and maximum practicable incorporation of crop residues or, better yet, adoption of conservation tillage methods. Reduced tillage practices – ranging from contour and strip-cropping, to no-till farming (where planting is done in a narrow slit without turning or stirring the remainder of the soil) – have a large number of environmental benefits. In erosion-prone areas they greatly reduce topsoil loss, and hence also lower nitrogen and pesticides transfer into sediments and surface waters. Their principal universal benefits are above all the reduction of overland runoff resulting in higher moisture retention and the improvement of soil structure.

In North America various forms of conservation tillage have moved from being isolated curiosities during the early 1970s to mainstream practices in many dry and erosion-prone areas during the 1980s. In 1989 32 per cent of corn, America's most important crop, was cultivated with conservation tillage methods, and another 25 per cent had between 15 and 30 per cent of the previous crop's residue on the surface (Conservation Technology Information Center 1989).

Although most of the research into efficient, low-input farming comes from the rich countries, many findings are readily transferrable to sub-tropical and tropical cultivation, and a number of traditional soil and water conservation practices should be studied and incorporated into modernizing agriculture (Reij 1991). Traditional practices should be also used as bases for improving the cultivation of local staple-food crops (rather than introducing one of the globally dominant grains), and for extensive dissemination and adoption of various agro-forestry arrangements.

Continuing reliance on local staples is especially important for the world's most rapidly growing populations in tropical Africa, where traditionally grown roots and tubers have numerous advantages over cereals (Pearce 1990). They fit better the high humidity climates, integrate very well with intercropping arrangements, can have very high yields, provide a higher yield stability, and some of them allow for flexible planting and harvesting. Growing crops and trees together lessen land competition, create desirable microclimates, provide additional nutrients to plants (above all when planting leguminous tree species shedding nitrogen-rich leaves and enriching the soil with rhizobia-fixed N) and, of course, trees and shrubs – inter-cropped, dispersed or adjacent, in shelterbelts or hedgerows – and yield continuous or cyclical harvests of food (fruit, nuts), fodder (leaves), fuel or timber (Nair and Fernandes 1984; Arnold 1990).

Traditional practices offer a large number of tree and shrub species suitable for all but the most extreme environments (Smil 1983). The commercial benefits of agrisilviculture may be quite impressive (Winter-bottom and Hazlewood 1987). Nigerian experiments found that hedgerows of *Leucaena* fixed about 100 kg N/ha every year, enough to support corn harvests of two t/ha. With fuelwood for cooking and heating farmers can recycle dung and crop residues rather than burn them, and crop yields can be increased by 10–20 per cent. And where grazing land is in short supply, supplementing a ration of grasses fed to stabled livestock with protein-rich leaves can greatly increase weight gains.

New food crops cannot make any notable global difference in less than one generation, but they can make a substantial local contribution in areas of pioneering introduction and improvement. The commercial potential of new tropical crops appears especially promising (Plotkin 1986). The most likely candidates for diffusion and breeding include amaranths (Andean grain with high protein content), the fruit of pupunha and buriti palms for eating, and those of pataua and babassu palms for edible oils. Over a longer term, most likely no sooner than in two generations, there is an even greater promise of crops bred by advancing techniques of genetic engineering (Tudge 1988).

Using water efficiently

Possibilities for a more efficient use of water resemble very much those of energy conservation. With more costly water reflecting better the real price of this critical resource, farmers, industries and households would discover a number of fairly inexpensive adjustments which would lead rapidly to considerable savings. And beyond these easy gains they would see a variety of more capital-intensive improvements whose gradual introduction would bring further long-term gains whose pay-back periods would be still very appealing.

Cropping is invariably by far the largest consumer of water in all areas with extensive irrigation, and improved efficiencies can free large volumes for urban and industrial uses. Potential gains are impressive (Stanhill 1986). Theoretical minima of evapotranspiration ratio (gram of water used for a gram of dry weight yield) for common field crops are between 320 for harvests in temperate climates and 800 in arid regions. In contrast, even the highly water-efficient Israeli cropping averages between 1000 and 1200, and the global mean for annual and permanent crops during the 1980s was above 5000. Consequently, there is roughly an order of magnitude difference between the minima and average world-wide performance, and a fivefold difference between this mean and the best national average.

As most of the poor world's irrigation water is delivered by traditional inefficient ridge-and-furrow applications, improvements can bring relatively high short-term rewards. Overall distribution, seepage and evaporation losses in the poor world's irrigation add up commonly to 50–60 per cent of carried water, and losses up to 80 per cent occur with simple furrow irrigation. Typical water use efficiency – the share of water released for irrigation that is finally evapotranspired by crops – is usually put at no more than 30 per cent even in the US surface irrigation, much below the rates obtainable with even relatively simple improvements, and far below the performance of advanced, but obviously more expensive, pressurized systems (Stanhill 1985). For example, in China a mere 10 per cent efficiency gain would release an equivalent of all tapwater supplied annually during the late 1980s to all of the country's cities (SSB 1991).

The most profitable gains are naturally those avoiding any new capital expenditures: irrigating every other furrow saves about one-third of water with only modest decline of crop yields. Careful scheduling of irrigation in accordance with the crop needs is the simplest low-cost measure reducing water waste. Irrigation modelling is now readily available in rich countries as a consulting service and microcomputers are turning it into an instant-response field tool. Optimal irrigation schedules, taking into account not only such variables as wind speeds and soil moisture, but also the appreciable differences in the sensitivity of various crops to water stress at different stages of their growth, can result in large water savings (Frederick 1988). This service may be relatively inexpensive even in poor countries when available to farmers in an area with fairly homogeneous soils and with only a few dominant cropping patterns.

A simple innovation introduced in the USA during the 1980s makes an almost perfect scheduling available to any farmer willing to invest only a very modest amount in an auger, a score of gypsum blocks and an AC resistance meter. Inexpensive gypsum blocks containing two electrodes are buried at several locations and depths in root zones (Richardson *et al.* 1989). Because the blocks absorb and lose moisture at a rate very similar to that of the surrounding soil, regular measurements of changing current flow with a pocket-size impedance meter gives reliable indications of the soil's moisture status. Tested benefits have included considerable savings for reduced water use and higher crop yields (as plants are much less stressed by avoiding either too much or too little moisture).

Better matching of crops with natural moisture supply is yet another highly effective alternative. For example, replacing corn with sorghum can lower the water need by 10–15 per cent, and planting sunflowers instead of soy beans as an oil crop can save easily 20–25 per cent of water. Where climate and market conditions are right the savings can come not only from reduced water use, but also from a higher profit from new crops. Such savings were realized by many Chinese farmers after the privatization of the early 1980s when they planted more oil and fibre crops and concentrated the water-intensive grain production on better soils (Smil 1993). But radical improvements of traditional irrigation efficiencies will be impossible without the gradual introduction of pressurized systems, including a variety of portable sprinklers, centre pivots, moving lines and drip systems.

Centre pivot irrigation systems spray water from a large number of emitters placed along pipes pivoting on wheeled towers around the point located at or near a tubewell and elevated 2 to 3 metres above the ground in order to ride above even the tallest crop (Splinter 1976; McKnight 1990). Lateral move systems pump water from a ditch running along the centre or a side of a rectangular field, and distribute it through long moving linear spray pipes. Both of these systems can irrigate moderate slopes and can cover very large fields. Centre pivot pipes are up to 400 metres long, covering a

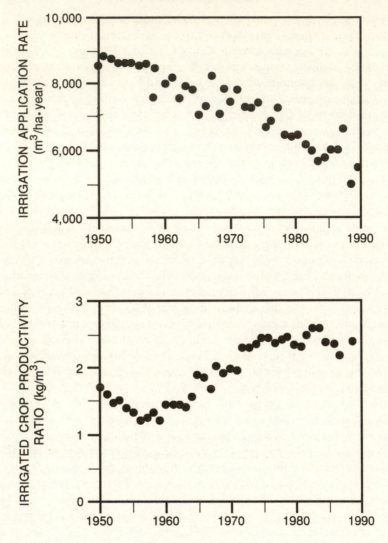

Figure 4.13 The figures show how the declining water application rates supported higher productivity of irrigated Israeli crops – but there has been a clear stagnation since the mid-1970s.
Source: Stanhill (1990b).

circle of 50 hectares, lateral distributors up to 800 metres long, irrigating rectangular fields of up to 150 hectares. Delivery can be programmed for optimum amounts of water, scheduled during night hours, and fertilizers and pesticides can be accurately distributed in the spray. Overall water use efficiencies average about 80 per cent, and may be as high as 95 per cent.

Drip irrigation systems distributing small volumes of water right to the roots of plants are considerably more expensive and are most suitable for growing high-value crops in arid climates. By 1990 about 60 per cent of Israeli irrigation was done with drip systems (Stanhill 1990b), and diffusion of this highly efficient (up to 95 per cent) method has been the main reason for the country's more than 30 per cent drop of average water-use rate and for the doubling of crop yield per unit of water application (Figure 4.13). And where the conditions are suitable, a combination of recent technical advances ranging from laser-guided land-levelling to runoff recovery systems makes it possible to raise the efficiencies of gravity systems close to those of pressurized techniques at a fraction of the cost required for the installation and operation of pivots, laterals or drips.

The total volume of water which can be saved world-wide by more efficient use in industries, services and households is only a fraction of the gains to be made by better irrigation, but national, regional and local benefits can be quite substantial. For many basic industrial processes these gains would be especially large in many poor countries. Their high water consumption rates are illustrated by the following comparisons of typical Chinese and western performances during the late 1980s. In China synthesis of 1 t of sulphuric acid needed more than 200 tonnes (t) of water (compared to the western mean of around 20 t), crude oil-refining consumed at least 5 to 6 t per t of feedstock (vs 0.2–1.2 t in the west), newsprint production used between 100-300 t of water per t of finished paper (no more than 50 t in the west) and steel-making needed two to three times as much water (and up to 20 times more of fresh make-up water) than in industrialized countries (Smil 1992b).

The poor world's industries are not only excessively high consumers of water, they also have very low recycling rates. Western industries have been recycling water for decades, but further substantial increases are possible: during the 1980s reuse rates for various American industries ranged between 5 and 20, but rates between 10 and 30 are practicable. Household per capita water use is commonly an order of magnitude higher in the rich countries than even in the poor world's cities, and so by far the most rewarding gains can be realized in North American, European or Australian homes. The most cost-effective approach is the installation of such inexpensive water-efficient fixtures as reduced-flow taps and showerheads. Retrofitting with better toilets and buying more efficient dish- and clothes-washers is considerably more expensive, but their purchase would be attractive for all new houseowners (Rocky Mountain Institute 1991).

In comparison with conventional fixtures, typical low-flow devices deliver the following water savings: taps close to 20 per cent, toilets just over 30 per cent, and showerheads around 40 per cent. Clothes-washers with water recycle can save nearly a third of water, and front-loaders about 40 per cent. Ultra-low-flow toilets and flow-limiting showerheads can cut the water by 60–90 per cent in comparison with standard fixtures. California's water shortages and the prospect of greatly reduced flow allocations have been a major driving force behind the commercialization of a large number of water-efficient household techniques. By 1991 American consumers could choose from nearly 100 types of water-efficient toilets, about 70 kinds of taps and 50 kinds of showerheads (Rocky Mountain Institute 1991).

Mass installations of such fixtures in Californian communities indicate the degree of rapidly achievable savings. In Goleta a combination of rebates for low-flow toilets and free distribution of 35,000 improved showerheads cut the per capita water use by 40 per cent in a year. And the economic viability of extensive household (and industrial) water conservation measures is notably improved by the fact that future reductions in water use will obviate construction of new waste-water treatment facilities.

As with energy and material conservation, there can be no simple generalizations, but the following gains can be seen as useful indicators of realistic possibilities. Low-cost measures which can be put in place in a matter of months or years and which would employ only well-proven techniques and easily available devices can cut water use in traditional irrigation by 10–15 per cent, in industries by at least 10–30 per cent, and in households by 20–40. Costlier approaches – irrigation sprinklers, high-performance fixtures and higher industrial water recycling – can cut water use by 40–60 per cent. The best available techniques, which may be cost-prohibitive in many instances, can bring water use reductions of the order of 60–80 per cent.

Reducing energy use

Long-term planning of energy use has been, almost without exception, biased toward the expansion of supply. Conservation, as Socolow (1977) noted, calls for an inverted emphasis, and also for an inverted methodological approach. Rather than starting with macroeconomic approximations and aggregate models, effective conservation strategies must focus on deliveries of particular kinds of energy. By examining in detail the existing supplies, conversions and uses, these approaches almost always uncover both some substantial rationalization opportunities and some better technical solutions.

Opportunities to save energy are everywhere: there is not a single class of major fuel and electricity conversion where technical and management changes would not improve the prevailing performance. Naturally, there are major differences in relative improvements and costs, and in the overall savings. Some converters, including large electricity generators and motors and large fossil fuel-fired boilers, have approached the limits of thermodynamic performance and marginal improvements would be expensive. In contrast, such common final uses as household lighting and heating continue to be surprisingly inefficient – yet better techniques are already highly cost-effective. Considerable energy savings are also possible by recycling and by expanding the use of materials with inherently lower energy costs.

By far the greatest impact would be achieved by improving the efficiency of the key fuel and electricity supply techniques and the performance of those relatively highly inefficient final conversions which account for substantial shares of overall energy consumption. This means different targets in different societies. In production these adjustments would range from improved extraction of hydrocarbons in old oil provinces of North America to extensive diffusion of natural gas-fired combined-cycle power plants in Asia. In households the focus would be on small coal stoves in China and oil-fired furnaces and refrigerators in the USA, in transportation it would be on trains and buses in India and on cars and trucks in Canada.

Especially large gains are possible when better converters become a part of a systematically improved design. When electricity generated in combined-cycle plants with overall efficiency of 50 per cent is converted to light in long-lasting fluorescent bulbs with efficiency of around 15 per cent, 7.5 per cent of the initial energy input ends up as a useful final flux; in contrast, only 1 per cent of fuel is converted to light when a standard power plant (about 35 per cent efficient) supplies incandescent light bulbs (3 per cent efficient). And high-efficiency furnaces in superinsulated houses may easily halve the total energy needed for heating in cold climates.

Such impressive savings may not be the norm in energy conservation, but when multiplied by 10^5–10^7 units then even marginal gains in conversion efficiencies of cars, refrigerators, washing machines, dryers, stoves and ranges add up to impressive nation-wide savings. Innovations will be needed, but surprisingly large improvements are achievable with widespread applications of existing techniques at acceptable costs: their adoption is primarily a matter of rational pricing and sensible legislation. In rich countries the efforts should be concentrated above all on residential heating and lighting and on car and truck performance.

Buildings are either the largest or the second largest consumers of energy (behind industries) in all rich countries. Space heating and cooling lead the demand, followed by lights and electric appliances and water heating. High conversion efficiencies, convenient use, easy maintenance and cleanliness, favour natural gas and electricity over solid and liquid fuels. The energy performance of buildings holds a tremendous potential for relatively inexpensive and lasting efficiency gains (Baird et al. 1984; Schipper et al. 1985; Rosenfeld and Hafemeister 1988; OTA 1991). North American means are particularly inefficient, but significant improvements are also possible not only in Europe but even in Japan. Actual per capita totals of primary energy delivered to households during the 1980s show national means ranging from just 11 GJ in Japan to 48 GJ in Canada (Schipper et al. 1985).

These large variations are explained by a combination of differences in climate (Canada averages about 4600 degree-days compared to Japan's 1800); preferred indoor temperatures (not uncommonly below 15°C in Britain, almost always above 18°C in North America); life styles (Canadians and Swedes heat between two and a half and three times as much water as the French or Japanese) and affluence (heating just a single room, or even a small part of it, with the family huddled around a Japanese *kotatsu* (small heater), compared to the central heating of America's sprawling bungalows). And although refrigerator ownership is now virtually universal in industrialized nations, there are major differences in the diffusion and use of such major appliances as clothes-dryers, dish-washers and freezers.

Proper construction is the single most important factor in reducing household energy use, but better furnaces, air conditioners, appliances and lighting are also needed – and are already available. Traditional furnaces operate with efficiencies ranging from well below 50 per cent for poorly maintained solid-fuelled units, to between 55 and 65 per cent for well-tuned traditional oil and gas furnaces with standing pilot lights. In contrast, the peak efficiencies for condensing non-vented natural gas-fired units with electronic ignition are between 91 and 96 per cent (Macriss 1983). Improvements in existing houses range from less than 10 per cent with simple insulation retrofits, to 25 per cent with a combination of more expensive retrofits (attic, basement walls, and double windows) and installation of better furnaces (70–80 per cent efficient), to 35 per cent with the best possible retrofits and the top furnace (efficiency over 90 per cent).

Conservation potential is well illustrated by comparing recent American improvements and possibilities against the overall mean, all expressed in kJ (kilojoules)/m^2/heating-degree day. Average energy use for the US housing stock was at about 160 in the 1980s – but buildings constructed during the 1980s averaged between 100 and 120, and superinsulated houses used only 30 to 50. The best non-commercial designs need as little as 15 to 20 kJ/m^2/ heating-degree day. Additional savings can come from structural changes. In comparison with a three-bedroom, single-storey house, an equally-sized two-storey building is 15 per cent, two-storey duplex 30 per cent, two-storey triplex 35 per cent and low-rise condominium apartment 40 per cent more energy-efficient (Burchell and Listokin 1982).

The energy performance of commercial buildings shows similar differences of national means and comparable potentials for future fuel and electricity savings. North America's glassy commercial high-rises built before the early 1970s need between 110 and 140 W/m^2 of floor area, or 2–2.5 kW (kilowatts)/m^2 of their foundations. Subsequent improvements were surprisingly rapid, and new buildings completed during the 1980s averaged commonly below 50 W/m^2, with many all-electric buildings using just around, or even below, 10 W/m^2. Most of the space heat in well-designed commercial high-rises is drawn from lights, office electronics and electric motors.

Individually, lights are overwhelmingly low-power converters, but combined savings resulting from high efficacy can be substantial. The most efficient incandescent light of the 1980s was a 10 kW source for film studios which produced 33.6 lm (lumens)/W (Weast 1990). Using 1.47 mW (milliwatts)/lm as the standard mechanical equivalent of light, even this powerful lamp is no more than 5 per cent efficient in converting electricity into light – while the rate for the common 100 W bulb is mere 2.6 per cent. In contrast, a standard 40 W fluorescent tube is almost 12 per cent efficient and it is also much more durable, lasting at least 25 times longer (20,000+ hours) than a 100 W bulb.

High-intensity discharge metal halide lights with efficacies of up to 110 lm/W, and high-pressure sodium lamps (maxima of 140 lm/W) are the most efficient commercial light sources, with efficiencies approaching or even surpassing 20 per cent. High-performance halide and sodium lamps are already in common use for outdoor lighting, and fluorescent lights dominate in the commercial market, but the substitution of incandescent household lights by more efficient sources has been quite slow. Further savings can come from a revision of lighting standards.

Since the nearly universal demise of coal-fired steam locomotives during the two decades following the Second World War (only China and India still have many of them in regular service), world transportation has been fuelled overwhelmingly by refined oil products. In 1990 almost 30 per cent of world-wide refinery output was motor gasoline, and just over one-third were middle distillates (BP 1991). Low shares of total fuel consumption and already high typical efficiencies mean that potential conservation gains would be rather small for shipping (powered by diesels and gas turbines), trains (powered by diesel or electric locomotives) and pipelines (powered by gas turbines). By far the greatest rewards can come from improving the performance of passenger cars.

Typical performances remain very low. The Otto cycle used by internal combustion engines has the highest practical efficiency of about 32 per cent, frictional losses lower the share to 26 per cent, and partial load factors (so common during city driving) reduce it further to about 20 per cent. Accessory equipment and automatic transmission may lower the total to as little as 10 to 12 per cent, and the actual performance in poorly maintained vehicles may be no higher than 7 to 8 per cent. In terms of actual fuel consumption, many cars of the early 1990s average 6 to 8 litres (l) of gasoline per 100 km even in city traffic. This is about half the rate of two decades ago, but far above the best commercially available subcompact and compact designs which need just 4.3–4.8 L/100 km when running at 90 km/h.

Much more substantial energy savings can come from a combination of gradual adjustments and widespread applications of existing techniques (Reitz 1985). Three design factors will make most of the difference: total car mass, engine and transmission performance and air resistance. Weight reductions cut power needs without any engine redesign: they come with front-wheel drive with transversely-mounted engines, and by replacing steel with lighter materials. Typical cars are still unnecessarily too powerful. Only when the driving calls for unusually rapid acceleration, travel on uncommonly steep roads or heavy towing is there a need to have passenger cars rated over 35 kW – yet even small cars are much more powerful. Lean-burn low-friction engines, continuously variable transmissions and greater diffusion of efficient diesels are other readily available options.

Extreme reduction of aerodynamic drag may become non-functional, but most vehicles are still far from that point. Earlier expectations assumed the best drag coefficient at 0.3, but the best production car in 1991 achieved 0.26, GM's Impact 0.19 and a Ford prototype 0.137, a better performance than an F-16 fighter. A number of other improvements, ranging from better bearings to smaller air conditioners will bring additional energy savings. These efforts should be most aggressive in the USA because the country uses a highly disproportional share of the world's total automotive fuel consumption. In 1990 US gasoline consumption of almost 330 Mt equalled 75 per cent of Japan's *total* primary energy use, and it amounted to nearly 5 per cent of the global fossil fuel conversions (BP 1991).

Opportunities for higher conversion efficiencies do not, of course, end with better buildings, lights and cars. High on any list of universally applicable choices must be co-generation, the utilization of waste heat from power plants and industrial enterprises requiring high-temperature steam or dry heat. This well-established engineering concept has been spreading since the 1970s (Hu 1983; Clark 1986). Another rewarding effort is improving the efficiencies of electric motors. They are fairly efficient machines but, given their huge numbers, even relatively small improvements can translate into big overall savings. Efficiency of small motors (below 1 kW) can be boosted from as little as 68–70 per cent to around 85 per cent, while large machines whose efficiency already exceeds 90 can gain just 1 to 2 per cent. High-efficiency motors have been available since the late 1970s and payback periods for their installation are just a few years.

And although air traffic is still only a minor consumer of liquid fuels in comparison to road transportation, world-wide consumption of jet fuel increased by nearly 50 per cent during the 1980s (UN 1980–91). Long-range projections foresee further substantial growth, above all in the Pacific region. Recent efficiency improvements have featured widespread adoption of turbofan engines mounted on wide-bodied jets (Sampl and Shank 1985). Even more efficient engines are about to enter commercial use.

Most of the conservation opportunities apply to both rich and poor countries: relatively low per capita energy consumption rates throughout the poor world do not imply limited opportunities for conservation. Inefficient conversions are pervasive both in the industrial sector operating many outdated plants, in transportation using aged and poorly maintained

EFFECTIVE STRATEGIES

vehicles, and in households wasting all but a small portion of biomass fuels in inefficient fireplaces and stoves. There are also notable opportunities for improving the performance of fossil fuel production: these measures ease environmental burdens by reducing the pressure for exploration and development of new resources. Given the fuel's global dominance, the most important technical challenge in fossil fuel extraction is to increase the output of existing oilfields.

For heavy oils the current recovery rates are below 10 per cent, for the lightest ones no more than 40 per cent, and the typical shares are between 25 and 35 per cent of the fuel in the parental rock. Moreover, many oil wells in older fields are only low-volume producers. For example, of some 650,000 wells in the USA about 450,000 produce less than three barrels of oil per day (less than half a tonne). Enhanced recovery techniques – including pressurization of reservoirs with water or natural gas, steam drive, and CO2 or surfactant flooding – produce more oil (usually between 10 and 30 per cent of the remaining oil-in-place) but are almost invariably expensive.

A recent innovation obviating any reservoir manipulation is thus of enormous potential importance: horizontal drilling of production wells not only increases output rate, but it also improves the overall recovery (Petzet 1988). This technique is especially effective in vertically fractured reservoirs. While standard vertical wells may either totally miss oil-bearing structures, or they can tap only a single fracture at a time, horizontal drilling may increase production up to five times.

Finally, at least a brief review of emerging innovations which can reduce environmental degradation either by converting fossil fuels with higher efficiencies or by replacing them. During the past two decades hardly any of the highly publicized alternatives to fossil fuels managed to make a substantial difference. Of course, this is not unexpected given the long waves of dominant primary energy modes. Consequently, a critical look at energy options before 2010-2015 (risking unapologetically to err on a conservative side) must exclude any commercial contributions of controlled fusion, fast breeder fission reactors, solar power satellites, ocean wave converters, giant windmills or kelp plantations. This still leaves a number of interesting alternatives. Foremost among them will be the continuing improvements of photovoltaic conversions (Hubbard 1989).

Annual world-wide installation rates of photovoltaic cells reached about 50 MW during the early 1990s, but the uses are still restricted to specialized applications, and conversion efficiencies are still too low. Fastest unit cost declines and greatest performance gains are expected to come from further improvements of amorphous silicon cells, from single-crystal silicon minispheres embedded in aluminum foil, and from an entirely new electrochemical cell absorbing radiation by photosensitive colloidal TiO_2 films (O'Regan and Gratzel 1991). Electricity generation by windmills has been among those alternatives whose future contributions were exaggerated during the 1970s, but a combination of technical improvements – above all aerodynamically efficient blades made with lighter materials and computerized controls – has lowered the costs of small- to mid-sized generators to make them a competitive option in some windy regions (Wilshire and Prose 1987).

But the best near-term practical prospects are for radically improved uses of fossil fuels. Given the fact that the global coal resources are so much more abundant and also much more equitably distributed than hydrocarbon deposits, introduction of clean coal conversions will be among the leading challenges of the coming decades. A particularly interesting innovation is using coal to power gas turbines. So far these power-generation systems have been using either natural gas or distillate oil, and it will not be easy to overcome the several inherent disadvantages of coal-based gas turbines: their higher capital cost, higher emission rates of both gaseous and particulate pollutants, and the increased corrosion and erosion of the turbine's metallic surfaces.

Nevertheless, research is progressing on both direct and indirect coal-fired gas turbines. Design goals are for commercial systems that will be 10–30 per cent cheaper than standard thermal power plants burning pulverized coal and equipped with flue-gas desulphurization, and because their efficiencies will be higher they would use 15–25 per cent less fuel – and hence also emit less CO_2 (Bajura and Webb 1991). Particulate and SO_x emissions would be further reduced with the use of fluidized-bed combustion, especially when commercial combustors – now with capacities between 20 and 150 MW – would be scaled up to at least 250 MW and could be installed in large power plants (Nelkin and Dellefield 1990). Limited capacities of fluidized-bed combustion can be also overcome by installation of multiple units (Smith 1991).

Conserving materials

Signs of wasteful excesses in affluent societies range from stuffing North American garbage dumps with massive editions of merely glanced-at newspapers (where news seems to be just an accidental presence among the flood of numbing advertising) to daily discarding of forests of disposable chopsticks and layers of elegant but almost obsessively overdone packaging in Japan (Hendry 1990). Even the least accessible places are not spared: a single Mount Everest expedition can transport as much 200 t of material and equipment to the base camp, and a large part of this mass is simply abandoned after the climb, creating a growing number of squalid dumps (Cullen 1986). And although most of the typical urban waste is made up of biodegradable materials, actual degradation rates in lined landfills are exceedingly slow: newspapers three or four decades old may be perfectly legible, and even vegetable remains may be largely preserved for a decade or two (Rathje 1989).

Extraordinarily wastefulness is relatively recent – waste dumping took off only during the 1950s in North America, and during the 1960s in Europe and Japan – but its combination with poor disposal practices and inadequate recycling is creating acute environmental problems for a growing share of affluent populations (Neal and Schubel 1987; O'Leary *et al.* 1988). American annual per capita garbage rates are highest for paper and plastics (over 250 kg and almost 60 kg vs less than 200 kg and less than 20 kg in Europe), but Western Europeans and Japanese generate about twice as much aluminum.

All of these wastes can be much reduced through a combination of better design, material substitution and recycling. Recent increases in recycling rates for metals, paper and synthetics can be further boosted by more rational pricing of waste disposal and by a fuller accounting of energy costs. The last measure is especially important because substantial fuel and electricity savings would be a great incentive for the recycling of all energy-intensive materials. A few examples illustrate these benefits. Steel-making from scrap will use only about 15 MJ/kg, 20–30 per cent less than the metal production from iron ore, while typical energy savings in recovering non-ferrous metals from scrap as opposed to their smelting from ores are at least 250 MJ (megajoules)/kg for aluminium, around 50 MJ/kg for copper and zinc and 30 MJ/kg for lead (Smil 1991a). Not surprisingly, aluminum has the highest recycling rates, but they are still only between 30 and 40 per cent in most countries.

New opportunities in metal recycling include the recovery of platinum group metals from catalytic converters in cars: the rates can be about 90 per cent for platinum and palladium and 80 per cent for rhodium (Frosch and Gallopoulos 1989). Although the recycled mass is small (less than 50 t of platinum-group metals are used as catalyst in a year), this recovery has enormous environmental benefits because it saves energy and avoids waste from processing ores containing only around 7 ppm of recoverable metals (seven kg of metal in 1000 t of ore).

Paper is the only non-metallic material which is commonly recycled. The highest national recycling shares are over 50 per cent in Japan and in several West European countries, but rates around 70–75 per cent should be the aim. De-inked pulp can be made with about 60 per cent less energy than the one from wood. Contrary to frequent claims that glass recycling saves no energy, it is possible to cut a few MJ/kg – but returning bottles for at least half a dozen refills is a much more rewarding option saving up to 16 MJ/kg. Benefits from recycling energy-intensive plastics are obvious but the hetero-geneity of wastes makes this task difficult. The best outlook is for recycling of mass-produced bulky wastes such as polystyrene cups and containers in fast-food eateries.

149

Material substitutions have been proceeding rapidly during the past two generations, but with more realistic resource pricing they could become even more effective in reducing both the initially invested energy and recyclable mass. Reducing the mass of energy-intensive materials is an option applicable to a vast range of products. Although unit gains may be modest, aggregates add up to major savings. For example, aluminum soft-drink cans averaged nearly 22 g (grams) in 1960 – but they weighed 17.7 g in 1990. British glass milk bottles went from 538 g before 1939 to 340 g in 1960 and 245 g by 1990, and plastic yoghurt containers from 12 g in 1965 to 5 g in 1990. Obviously, these reductions are limited by material properties, but there is still much scope for further improvements.

Extensive conservation of many materials essential for the development of industrial civilization will be greatly helped by the fact that their markets throughout the rich world are saturated and their relative importance in national economies has been declining (Larson *et al.* 1986). For example, the rate of the US steel consumption per unit of GDP reached its peak in about 1920, and it has since declined by more than one-half. Other declines have been much more rapid: American use of copper per unit of GDP expressed in constant monies fell by almost 60 per cent between 1970 and 1990 (US Department of Commerce 1991).

Other rich economies, above all Japan and France, have never reached the peak levels comparable to US rates. And recent developments in China indicate that the declines of material use per unit of GDP can start much earlier in today's industrializing economies and that they can be much steeper. Consequently, it would be very misleading to present historic trends of average material consumption rates for rich nations as the models for industrializing countries. Many poor countries are still very much in the stage of ascending rates of material consumption, but most of their peak utilization levels will be well below the past western experience, and, by taking advantage of new technical advances, will be able to decline faster.

These trends make it unlikely that the land-use impacts of extracting basic metals and building materials will keep on expanding at rates prevailing during the latter half of the twentieth century. But some material substitutions may have greater environmental impacts in spite of the greatly reduced mass: for example, displacing of iron and copper by aluminum and silicon requires considerably more energy for producing the two lighter elements. While modern blast furnaces produce iron from ore with 20–25 MJ/kg and while copper smelting needs between 60 and 125 MJ/kg of hot metal, aluminum from bauxite costs between 220 and 350 MJ/kg, and making silicon from silica uses about 230 MJ/kg (Smil 1991a).

But given the relatively small consumption rates of these elements (even in the US they are just around 20 kg/capita for aluminium and less than 3 kg/capita for silicon), high energy costs are only a small fraction of overall energy use (typically less than 5 per cent), and they will almost certainly be reduced by new production techniques, and in the longer run they should be of even less concern in the deployment of efficient solar conversions.

Looking beyond the next generation, one of the most helpful advances would be practical large-scale enzymatic hydrolysis of cellulosic wastes. Energy-rich cellulosic wastes – newspapers, cardboard, junk mail, old clothes, disposable diapers, vegetables, grass clippings and tree leaves – now account typically for 40–50 per cent of all household garbage in rich countries. Although biodegradable, these wastes decay very slowly. Composting could greatly reduce the volume of these wastes, and potential markets appear to dwarf production capacities (Gillis 1992).

But a better future solution would combine a substantial reduction of solid wastes with generation of high-quality liquid fuels by using cellulase enzymes capable of catalysing the breakdown of the polysaccharide into glucose which can be turned into ethanol. These enzymes are produced by a number of celullolytic fungi, above all by the genus *Trichoderma*. Harnessing this ability on an industrial scale will require not only high-volume production of requisite enzymes, but also their more efficient use and recovery and reuse after the completion of hydrolysis, as well as the development of cellulase components with enhanced specific activity reducing the amount of required enzymes (Woodward 1991).

Even this brief review of either already commercial or promising techniques is sufficient to indicate the richness of opportunities available to minimize environmental degradation. But this promise, as well as the effectiveness of pricing and management innovations and life-style adjustments, will be circumscribed, and sometimes largely negated, by a large number of complications facing the widespread implementation of both technical and socio-economic changes.

5

CASCADING COMPLICATIONS

EMBEDDED LIMITS OF ENVIRONMENTAL MANAGEMENT

Fortune does not arrive in pairs, trouble does not come singly.

Chinese proverb

Having the tools, technical and managerial, and being aware of effective solutions, is only the beginning of challenging transformations. In the case of environmental management, translating the promise of possibilities into the reality of accomplishments is especially complicated. The interconnectedness of biospheric processes guarantees that environmental changes always embrace a multitude of states and processes, and effective solutions must pay attention to these inherent complexities. Inevitably this slows down their rate of diffusion and limits their reach.

Moreover, even our best-intentioned actions engender new problems which, in turn, give rise to further difficulties, ranging from marginal annoyances to challenges more intractable than the original complication. These unexpected, often counterintuitive, consequences further expand the limits on the modes, efficacies, costs and rates of our responses. No ranking of all such limitations makes sense, but among the most important ones are the continuing high rates of population growth in most poor countries, combined with the desire for strong economic expansion, the necessity to design our remedial actions in accord with natural constants, and the desirability of anticipating undesirable environmental feedbacks. No less important is the realization of the often limited power of seemingly outstanding technical fixes and the complex appreciation of enormous economic and social adjustments arising from our efforts to do things differently.

And even before we get to choose a course of action we must first deal with many fundamental uncertainties surrounding the understanding of our intents and goals. How do we define adaptation, what do we really mean by sustainability? Is the quest for monetary valuation of environmental goods and services a necessarily difficult but an eventually achievable goal

152

– or is it merely an intellectual mirage? And, of course, there is little we can do about the counterintuitive, unforeseeable consequences of our best-intentioned efforts.

Progressing through the following coverage of complications and obstacles on our way to a more enlightened management of environmental affairs will be an exasperating experience: often it may seem that I am dismantling previously proffered solutions with such a zest that I am undermining their basic credibility. I believe this approach is absolutely necessary: it goes beyond playing an *agent provocateur* (in itself a desirable effort) as it leads us to confront the enormity of required technical fixes and, even more so, of necessary social adaptations. I firmly believe that the challenge is surmountable – but not unless we face its true magnitude.

FUNDAMENTAL UNCERTAINTIES

Nature which governs the whole will soon change all things which thou seest, and out of their substance will make other things, and again other things from the substance of them, in order that the world may be ever new.

Marcus Aurelius, *Meditations*

This Aurelian observation should be kept in mind by all students and would-be-managers of environmental change. All too often I have the feeling that, in spite of their self-professed ecological pedigrees, many environmentalists have an anti-evolutionary mentality in desiring to preserve the unpreservable status quo. Inevitability of change always makes all projections, forecasts and plans risky. External manifestations may matter very little: unobserved – either because initially unobservable or because unlooked for – changes are frequently happening rapidly behind the façade of deceptive constancy.

Such unforeseen changes leading to unforeseeable outcomes will be always with us, but they are not the only important category of fundamental uncertainties complicating effective environmental management. Ubiquity of change means that all normative approaches to environmental manage-ment must be suspect: adaptation is a constant challenge, not a finite progression toward a clear goal. And if the ultimate goal is sustainability, then the paths to its achievement will be radically different depending on the content of that fashionable, but practically so elusive a concept. I cannot foresee an adoption of a universally acceptable definition of sustainability because the idea is inseparably connected with complex definitions of needs, rights and obligations.

And although the most effective way toward more efficient, more responsible, more rational society, which would be the best guarantee of sensible environmental management, is a radically improved valuation of all goods and services, I cannot foresee how this desirable quest could be rid of fundamental uncertainties either. More inclusive pricing and more realistic national economic accounts will always contain notable shares of assumed valuations, and will not be able to internalize satisfactorily large segments of environmental costs because of insurmountable difficulties in converting such losses into monetary equivalents.

Adaptation and sustainability

The great evolutionary goal can be stated so simply: industrial civilization has to adapt in order to become sustainable. The key words – adaptation and sustainability – seem so comprehensible to the modern mind, so intuitively appealing. In reality both of them are ill-defined terms, some may argue even indefinable, except in narrowly specific or excessively cumbersome ways. This lack of precision matters: we design our actions to fit those words and inadequate definitions may lead to ineffective outcomes. Common understanding of adaptation offers no useful guidance for our future actions. Its insurmountable weakness is revealed by simply looking at how we judge successful evolutionary adjustments: by identifying the assorted idiosyncrasies of the survivors (the best adapted or the fittest organisms) and then equating them with the adaptations conferring the highest fitness (Bethell 1976).

Obviously, 'this procedure still relies on the empty claim that the fittest . . . survive better' (Peters 1991), linking evolution and adaptation in an insolvable tautology. So it is perhaps not so surprising that modern ecological syntheses have little to say about adaptation. The term did not even make a single index entry in E.P. Odum's (1971b) *Fundamentals of Ecology* or in *Ecoscience* by Ehrlich *et al.* (1977), and it gets just a passing mention in H.T. Odum's (1983) *Systems Ecology* and in *Global Ecology* (Southwick 1985). And it is placed only 12th among the 25 most popular ecological concepts identified by a poll of the members of the British Ecological Society (Cherrett 1988).

Even when putting the evolution-adaptation tautology aside, understanding adaptation remains elusive. Complexities of organismic and civilizational adjustments fit Brooks and Wiley's (1986) general conclusion that 'some things cited as causal agents of evolution are actually effects of other causal agents. This is true for ecological interactions in particular'. And Endler and McLellan's (1988) admission that 'a large number of processes cause and direct evolution, but we know virtually nothing about most of them' also clearly applies to the growing complexity of human societies and to their adaptations.

Two fundamental problems determine the conclusions regarding the success or failure of adaptive processes, and both stem from the impossibility of defining successful adaptations *a priori*. First, is it the optimal fit or merely some degree of improvement defined according to the available time and genetic and environmental constraints? And what position on the continuum ranging from adaptations of all traits at all times or places to some traits being adaptive sometimes in some places will be good enough to confer, *a posteriori*, the mark of successful adaptation?

Second, what is the environment that organisms are trying to fit in improved ways? If environments remain approximately constant for long periods of time then it is possible for the adjustments to move the affected populations to (or at least closer to) new equilibria. But what if the changes are much faster than any practical rate of adaptation or, even more importantly, what if there is no practically achievable equilibrium to aim at? Given the often very rapid rate of change we may be repeatedly trying to adapt to conditions whose gravity will be greatly surpassed by the intervening developments – or we may adopt unnecessarily drastic measures whose eventual completion could be clearly counterproductive.

Another fundamental challenge arises from the shifting relations between parts and whole in human evolution. Human evolutionary success has been primarily due to generalized adaptability, not to narrow specialization. But this heritage is in contrast with ever more specialized demands of modern industrial civilization, and we have no way of assessing prospects of successful adaptation based on the apparent challenges of the task and the known capacities for change.

To the great surprise of biologists the finches of the Galapagos Islands, the birds which gave Charles Darwin perhaps the best-known example of how natural selection works over millions of years, have been recently observed to evolve in real time (Grant 1991). A single drought can change a finch population characterized by beak depth, a trait determining the ability to crack large seeds produced during the spells of low rainfall. Before extensive and repeated capturing of finches in mist nets we had no idea that the adaptation can be so rapid.

In contrast, we are still puzzled by the sudden demise of many human societies, including those which have not only adapted to specific environmental conditions, but which have achieved admirable levels of complexity, including intensive farming and effective social organization capable of coping with some formidable external challenges. Fascinating analyses and theories of collapse abound (Tainter 1988; Yoffee and Cowgill 1988), but we still do not know why such seemingly successful societies as the Harappan civilization, the Hittite or Han empires or the classic Maya disintegrated so rapidly.

If adaptation is the process by which organisms and populations provide better solutions to problems of environmental change, then it cannot be obvious what caused the eventual failure to adapt in human societies. Was the environmental change either so rapid or so extensive (or both) that the affected society, although doing its best and surpassing its former creativity and management skills, simply was not able to cope? Or were the external changes rather ordinary, equal to challenges successfully met before, and was the failure to adapt rather a matter of an utterly inadequate response falling far below the previous achievements for the affected society? If so, what were the factors responsible for the failure to formulate and to execute an effective response: an epidemic, profound social malaise, a violent conflict?

Amassing book-length arguments about the inexhaustible creativity of the human spirit or about the inevitable decline and eventual collapse of modern industrial civilization is easy – and not very helpful. We simply do not know if we have already caused too much irreversible environmental damage, or if we will be able to summon again and again the necessary wisdom and will to act in order to adapt to what are still apparently manageable changes. This fundamental uncertainty alone would make all discussions of sustainability highly suspect. A closer look reveals that even if we assume the best, translating this deceptive concept into realistic action would be immensely difficult.

The idea of sustainable development was first popularized by the International Union for the Conservation of Nature and Natural Resources (IUCN) in its 1980 proposal of the world conservation strategy. IUCN did not explicitly define the goals of this strategy, it merely stated that the approach 'must take account of social and ecological factors, as well as economic ones; of the living and non-living resource base; and of the long term as well as short term advantages and disadvantages of alternative actions'. In contrast, the widely cited definition offered by the World Commission on Environment and Development (1987) specifies that sustainable development should meet 'the needs of the present without compromising the ability of further generations to meet their own needs'.

Contentious meanings of sustainability have generated plenty of writings probing the concept both as a theoretical construct and as an attainable goal, and critical reviews of this work are available in Barbier (1987), Redclift (1987), Simon (1989) and Lele (1991). To me, problems of scale and the relationships between parts and wholes come immediately to mind, and a fundamental existential reality is their perfect illustration.

A strictly organic, self-contained farming (animal draught, no fossil fuels) with a part of the land set aside for growing trees (for fuel and timber and tools), and with the only external energy and material subsidies received as simple machines and tools is obviously sustainable on the civilizational time-scale (10^3 years): it was prevalent everywhere during the pre-industrial era, and it can be seen even today (with somewhat greater external influences), not only in isolated parts of Asia, but also in North America, where a declining number of the Old Order Amish farmers does not use either tractors or electricity (Stinner *et al.* 1989).

But this eminently sustainable way of farming can support typically no more than ten people per ha of farmland, and it requires participation of more than 80 per cent of all the available labour force. Consequently, it is not an option for any densely settled poor populous nation unless it would be first able to rid itself of a major fraction of its people (even an instant zero population growth would not be good enough) – and it could be practised in land-rich industrialized countries only when about nine-tenths of their inhabitants would be willing to leave the cities and to accept a drop in their standard of living to levels prevailing three to five generations ago.

Simply put, today's societies – be they affluent, highly urbanized and farmland-rich, or poor, predominantly rural and with little arable land – cannot return to purely solar agriculture and retain their existing socio-economic arrangements. But they will not be able to sustain the prevailing farming practices for thousands of years into the future either. Modern farming is simply too dependent on material and machine inputs energized largely by fossil fuels, puts too much stress on animal foods and is ecosystemically too degrading to be sustainable over a period matching the run of traditional cropping (at least 5000 years).

A call for sustainable farming is thus an unprecedented challenge to create a new system, an arrangement whose particulars cannot be designed and put into practice and whose durability cannot be estimated without knowing first the likely trend of principal external factors and key output specifications. Its possibilities will be determined both by population totals and by average consumption levels and environmental constants.

Openness of living systems makes it enormously difficult to determine when sustainability ceases to be only a matter of endogenous relationships and when it becomes so dependent on external factors that the system level should be defined more broadly. In time of rapid global changes even otherwise self-contained and sustainable agro-ecosystems would be vulnerable to stratospheric degradation and to shifts in climate patterns. Consequently, possibilities of global climatic change force us to define sustainability at the highest system level – the biosphere – and then to proceed downwards. Lynam and Herdt (1988) are certainly correct to see this proposition as unsettling: it makes everything so much harder, because the future course of global changes is so uncertain.

But even if we would know exactly what levels of pollutants, or what mean temperatures we would be facing during the next several hundred years, the durability of new arrangements would still depend primarily on output levels. At the highest output levels the global sustainability horizon disappears instantly: extension of the North American food pattern (per capita meat consumption averaging about 110 kg/year, 40 per cent of food energy supplied by fats) to the rest of today's humanity is physically impossible. In 1990 the global population eating meat at the US rate would have needed 580 Mt of it, nearly 3.5 times the world-wide output. This level of production would require the feeding of between 55 and 70 per cent of

157

the recent world-wide grain harvest to livestock (USDA 1991), an impossibly high share in all of the populous poor nations where 80–98 per cent of grain is consumed directly as food – and even so it can meet at best only the basic existential needs.

Adoption of the Japanese norm – implying average annual landings of fish and other aquatic species at just over 70 kg per capita – is also impossible. The global aquatic catch would have to go up more than fourfold, yet many traditional fisheries are experiencing declining yields (World Resources Institute 1992). Something between the Chinese and Japanese diet would work well for today's global population – but we really do not know how sustainable are the high applications of synthetic fertilizers required by such a relatively restrained consumption mode. Analogically, different sustainability designs could be constructed for various populations over arbitrarily chosen periods for consumption of basic minerals, wood or hydroelectricity.

Time-scales matter enormously. Is the idea of a civilization surviving as long as blue-green algae (10^9 years) more than an intriguing thought? Would it be practically almost as forbidding to match the evolutionary span of terrestrial mammals (10^7 years)? But even if we would aspire to nothing more than matching our stay as a sapient, high-culture species – that is, striving to be around for at least another 10^3–10^4 years – then it is not at all certain what path to follow. Do we put less stress on intergenerational equity and try to optimize our welfare and that of our immediate successors? Or are we putting world-wide equity, including the intergenerational parity, first, and are we trying to bring the minimum acceptable existential benefits to the largest number of people over that time?

Strict conservationists favour the second path, but to traditional economists 'the argument that a moral obligation to future generations demands special treatment of environmental investments is fatuous' (Summers 1992). They believe that our descendants can be helped as much by improving infrastructure as by preserving rain forests. Or are we trying to maximize the quality of life for a smaller number of people by drastically reducing the future global population? Perhaps the only thing we can state with confidence is what sustainability is not.

While assessing the prospects of sustainable agriculture, Wilken (1991) writes flatly that 'no system can be considered sustainable if it cannot accommodate increased numbers and increased demand'. In this he is echoing the standard economic understanding: sustainable development meaning 'the sustainable growth of per capita real incomes over time' (Pearce 1989). Indeed, the term sustainable growth can be seen in recent writings almost as frequently as sustainable development – but such a phenomenon is, of course, impossible on a finite planet. Daly (1991) is right when he calls sustainable growth 'a bad oxymoron – self-contradictory as prose, and unevocative as poetry'.

As it grows, the global economy, an open subsystem of the finite biosphere, incorporates an ever larger share of the Earth's ecosystems. Vitousek *et al.* (1986) estimated that in the early 1980s about a quarter of potential global net photosynthetic production was appropriated by human societies, and that the share was around 40 per cent for terrestrial photosynthesis. This may be an exaggerated estimate – but the process has clearly obvious limits. Continued growth is impossible, but continued development, a term implying always a change but not necessarily growth, is not only possible but highly desirable, both individually and collectively.

This reality leads to yet another set of challenges, forcing our industrial civilization, a creation of rapid economic expansion, to come to terms with prospects of a society where progress is not measured by growth of material and energy indicators. This would be an extraordinarily daunting transformation: just recall how rattled even the most affluent countries get during brief recessionary spells when the GDP growth is only a fraction of a per cent, or when there is even a temporary decline in total economic activity. These spells are seen as regrettable aberrations, and everybody clamours for the resumption of growth, the expected state of modern societies. On an individual level, this shift would require delinking of social status from material consumption, a no less revolutionary shift than the acceptance of no-growth economies.

This unsustainable craving may be a flawed, although generally advocated, economics – but it has also become an essential formative force of modern societies. Doing away with it would change our lives profoundly, in ways both predictable and unforeseeable. Environmental gains of zero-growth economies would be bought by considerable social and economic sacrifices and this burden would be especially heavy for the first generation making such a transition after centuries of (interrupted but then always resumed) growth. The most painful consequence could be the freezing of prevailing levels of affluence or poverty leading to acute social instability.

Historical evidence confirms that people can endure poverty and physical hardship as long as their efforts keep on yielding noticeable improvements; if their struggle brings hardly any rewards, the chances of violence go up substantially. This led McKean (1973) to note that 'maybe growth will appear to be like democracy: the worst possible situation one can imagine – except for the alternatives'. The only hope to improve the lives of poor people in a no-growth global economy would then rest on an unprecedented altruism of the rich, and this massive redistribution of wealth would have to be going on both intra- and internationally. How strong is the constituency favouring such a deal among affluent North American and European populations – especially when one considers that even the approach of a stationary state and the retraction of net investment in market economies could result in crises akin to the Great Depression of 1929–32 (Boulding 1973)?

Zero growth would also largely eliminate one of the mainsprings of western progress, the quest for social and economic mobility. Are we ready for this? Such questions slip the matters of sustainability into the realm of social acceptability and political support – and no indication of the concept's fate will be clearer than our willingness to pay the price for maintaining the biospheric integrity. But the task of setting the prices right is beset with so many fundamental uncertainties that only a naïve mind could expect a rapid progress in this critical direction.

Mirages of valuation

Exhortation and regulation are not going to result in sufficiently rapid declines of energy, water and mineral consumption, and continuation of current accounting practices, camouflaging declines of natural assets as income generation, will not be able to preserve critical environmental services and irreplaceable biodiversity. We need to do much better, but both the quest for real prices and the effort to rationalize national production accounts will be extraordinarily difficult. Because I illustrated the need for more inclusive pricing by reviewing the externalities of energy supply, I will use this example again to outline the challenges of internalization.

Although we should not underestimate the progress made toward the fuller pricing of energy (or its convertors) during the past two generations, we must realize that most of the challenge is still ahead – and that it will be most difficult to internalize precisely those burdens which appear to have the highest external costs. The intertwined problems of scale (what to include) and complexity (how to account for it) are the two principal, and often intractable, obstacles. Scale problems are both spatial and temporal, and could be seen predominantly as a matter of boundaries. Unequivocal selection of a proper analytical set is often impossible even when the requisite monetary accounts are readily available.

Perhaps the best illustration of this challenge is to try to estimate the real cost of the Persian Gulf oil to American consumers. This concern, made so obvious by the Iraqi invasion of Kuwait and the subsequent American response, is also a perfect example of extended linkages which must be considered when dealing with environmental change: assuring the flow of low-priced Middle Eastern oil by investing heavily in the military protection of the region is clearly one of the most important reasons for excessive air pollution from burning gasoline, diesel and fuel oils. To protect this supply the USA led a large military alliance, first in mounting armed deployment in the area (Operation Desert Shield, August 1991–January 1992), and then in a bombing campaign and the ground war (Operation Desert Storm, January–March 1992).

The US Department of Defense estimated the total funding requirement for these operations at US $47.1 billion, and total incremental cost at US $61.1 billion. But the General Accounting Office pointed out that the armed services accounted for the total costs without any adjustment for the expenditures they would have to bear even in the absence of the Persian Gulf conflict (GAO 1991b). Consequently, GAO believes that the DoD's estimate of incremental costs is too high – but it estimates that the total cost of the operation was actually over US $100 billion.

Even if we would assume that the whole cost of the Desert Shield/Storm should be charged against the price of the Persian Gulf oil, which sum are we to choose? But such assumption is hardly realistic: Iraq's quest for nuclear and other non-conventional weapons with which the country could dominate and destabilize the region, and implications of this shift for the security of the US allies certainly mattered as well. But there is simply no objective procedure to separate the costs of such a multi-objective operation according to its varied but interconnected goals. The operation's costs to the USA are even less clear because of substantial burden-sharing: total pledges of US $48.3 billion of foreign help were almost US $800 million above the Office of Management and Budget's estimate of US funding requirements. One could stay on these narrow accounting grounds and argue that the country actually made a profit!

And whatever the actual cost of the operation to the US, against what amount of oil is this to be pro-rated? Merely against the US imports of the Persian Gulf crude (recently less than one-fifth of all US imports) or against all shipments from the region (they go mostly to Japan and Europe) or, because a stable Gulf means a stable Organization of Petroleum Exporting Countries (OPEC) and promotes low world oil prices, against the global production? And what would be the time-scale for pro-rating the costs? Certainly not just the months of the operation, but who knows for how many years will that intervention stabilize the region? And with continuous stationing of US forces, weapons and supplies in the Gulf, there will be further incremental costs. And the past price was not cheap either: the USA have maintained a naval presence in the Gulf since 1949, and it was augmented by lengthy displays of British and French forces (GAO 1990).

What kinds of the Navy's costs should be counted: totals needed to create and to operate the task forces, or only the incremental outlays; only the ships and the planes in the Gulf, or also those of the nearby Indian Ocean and the Mediterranean fleets? And is it correct to leave out from the real price of the Persian Gulf oil the costs of US economic assistance to the countries of the region? Between 1962 and 1990 US net disbursement to Israel, Egypt and Jordan added up to almost US $80 billion, with about three-fifths of this total going to Israel (US Department of Commerce 1960–91). Should not the propping up of the two friendly Arab regimes and the strengthening of the country's principal strategic ally in the region be seen as yet another cost of Middle Eastern oil? Obviously there can be no non-judgemental agreement about what share of that aid should be charged against the cost of crude oil exported from the region.

But once we would agree what to include and how to standardize our accounting we could calculate the military or foreign aid costs of the Gulf oil with acceptable margins of error. The same cannot be said about the environmental and health costs of energy conversions. In most cases we are not only unsure about what to include, but the complexity of impacts makes it also very unlikely that we will be able to express their costs with high confidence. Attempts to estimate the costs of air pollution, or to value the benefits of clean air, are perhaps the best examples of these challenges. In spite of decades of research there are no fundamentally new conclusions offering a clear guidance in assessing the long-term effects of air pollutants on health.

Numerous statistical analyses seeking to unravel the relationship between air pollution and mortality have been handicapped by dubious design, interferences of unaccounted variables, and the difficulties of separating the effects of individual pollutants when people are exposed to complex, and varying, mixtures of harmful substances. Ferris's (1969) conclusion is still true: general mortality statistics appear to be too insensitive to estimate effects of chronic low-level air pollution with accuracy. Consequently, major differences in calculations of health costs of pollution (or health benefits of abatement) continue. In the early 1980s extreme estimates of costs in a major study on the burdens and benefits of SO_2 controls in the Organization for Economic Co-operation and Development's (OECD's) European nations ranged just twofold for corrosion and 2.8fold for crops, but the difference was 25fold for health effects (OECD 1981)!

A study sponsored by the California Air Resources Board estimated annual mortality and morbidity benefits from meeting the air quality standards in the Los Angeles Basin at between US $2.4–6.4 billion (Rowe *et al.* 1987), while a 1989 assessment prepared for the South Coast authorities offered the best conservative estimate of US $9.4 billion a year, with a plausible range between US $5 and 20 billion (Hall *et al.* 1992). Continued large differences between extremes – a more than eightfold discrepancy in the two California cases – makes cost-benefit studies of air pollution control a matter of unceasing scientific controversies (Krupnick and Portney 1991; Lipfert *et al.* 1991).

Evaluating the chronic health costs of any environmental degradation will always be highly disputable because of a large number of factors shaping human well-being and survival. These considerations include not only nutrition but also many hard-to-quantify social circumstances: disease rates increase where supportive ties between people are interrupted, where people occupy low positions in a hierarchy, and where they are disconnected from their past (Lindheim and Syme 1983). How can we ever separate these effects from those of chronic low-level environmental exposures?

And yet these uncertainties seem relatively small in comparison with challenges inherent in valuing biodiversity. Only a few species we wish to preserve have a clearly established commodity value, a very large number of them have obvious amenity value (pleasure or excitement of seeing migrating geese, hovering hummingbirds or sequoia groves), and every one is undoubtedly entitled to a moral value, independent of any of our uses or feelings. At present we cannot quantify values in the latter two categories, but even if we could do so this divide-and-appraise method would be ecosystemically wrong: species do not exist independently, they have co-evolved with their communities. Moreover, the divide-and-appraise method does not even ask the right question, because the value of biodiversity is definitely more than the sum of its parts (Norton 1986).

But in practice an uncompromising approach of infinite valuation will get us nowhere. Ehrenfeld (1986) notes that although we do not know how many plant species are necessary to perpetuate productive forests, we could see that even the demise of such a dominant species as the American chestnut has not brought the eastern deciduous forest down with it. 'Who knows what genetic resources exist in the unexplored DNA of a silverspot butterfly?' asks the Vice-President of the Wilderness Society (Shaffer 1992). To argue that every butterfly, tropical liana or obscure tuber is literally beyond pricing because its extinction will mean the permanent loss of genetic information assembled by a long evolutionary span is actually an abdication of rational decision-making responsibility.

Only slightly less unrealistic is to believe that every threatened species may contain as yet unidentified compounds capable of miraculous feats of healing, feeding or fuel production, and hence it could be potentially of enormous economic value. Because it will be clearly impossible to save every species threatened with extinction, we should be devising long-term priority conservation plans. Such a strategy is preferable to species disappearance by default – but we have no objective ways of ranking the threatened species, and then deciding what share of available resources we should devote to the preservation of species with identical or very close priority ranking.

How badly such attempts can work is best illustrated by the experience with the American Endangered Species Act of 1973 which imposes fines or prison sentences for harming and capturing species on the steadily growing list maintained by the US Fish and Wildlife Service. Priority rankings on this list might as well be arbitrary because the government spending on species recovery shows no correlation with its ranking (Mann and Plummer 1992). In 1990 more than half of US $100 million spent nation-wide on all endangered biota went to less than 2 per cent of the species on the list. The Act protects primarily high-profile individual species rather than overall biodiversity (Rohlf 1991).

This tilt is most evident toward 'charismatic megafauna', the category perhaps best represented by California condors and Florida panthers: it received 14 times as much funding as other species. But what is least explicable is that species whose survival is threatened received on the average more funding than those who are already endangered! If this is how the world's richest country with an unusually powerful species protection Act goes about the business of conserving biodiversity, how can anybody criticize Peru's or Zaire's actions – or hope that we can soon bring some rationality into long-term conservation efforts?

Some of the new valuation approaches advocated by environmental economists look intriguing but their general adoption is unlikely. For example, it is true that non-use values may be much higher than use values: few people will actually travel to a remote national park but millions who see its images may be willing to pay for its continued preservation. Applications of this contingent valuation, or willingness to pay, have been spreading. All that is required is to ask a representative sample of the concerned population how much they would be willing to contribute each year to have the amenity preserved, or even enhanced.

Final sums obtained through contingent valuation can easily surpass the economic benefits of any contemplated development (coal-mining, river damming) – but the process is inherently flawed. The basic question is: would the same public opinion survey elicit comparable levels of willingness to pay if the question would not be the preservation vs the development of one site but, more realistically, the preservation of scores or hundreds of threatened environments in a country vs the nation's economic growth? Naturally, the same question applies to contingent value of individual threatened species. People in rich countries may be willing to pay US $50/household to see African elephants preserved, but would they extend this contingent valuation to thousands of threatened species, including scores of beetles which even entomologists have a hard time to identify?

This example underscores a fundamental obstacle to the full pricing of commodities responsible for environmental degradation: there are no markets whatsoever for obscure miniature beetles – or for a ClO_x-free stratosphere above Antarctica, and only limited and imperfect ones for some intact ecosystems (in national parks and game reserves). In the absence of such markets, and knowing they will not emerge spontaneously, our quest for more inclusive pricing will have to rely on the use of social prices, that is, on externally supplied valuations trying to assess the worth of environmental resources for the long-term well-being of affected societies or, indeed, of mankind as a whole.

This returns the challenge into the domain of political acceptability, exposing it to the perils of short-term economic and electoral expediencies, and making it easier to argue for simple taxation. Actually, there is little in recent proposals for full pricing of commodities by using assorted pollution or inefficiency taxes which Pigou (1920) did not think about, and has not tried to resolve by proposing that the tax paid by the offending party should equal the marginal external cost incurred by the affected parties. This would provide an excellent mechanism for internalizing any intolerable externalities.

Except, of course, that calculating such a tax would not be any easier than establishing the full market cost of the commodity whose supply is causing environmental degradation, or is otherwise draining limited economic resources (as is so demonstrably the case with the US Middle Eastern military involvement). There is simply no way of making the critical decisions about the worth of environmental goods and services without a recourse to political and social compromises. Effective policies must balance a variety of interests – including industrial viability, political acceptability and regional interests – which have legitimate stakes in any modern economy (Skea 1992).

A perfect Pigouvian carbon tax dictated by fears of global warming could mean the end of the coal industry. A small European country with marginal coal production may accept that verdict – but could China, the nation with one-fifth of the world's people, a third of known coal reserves and nearly three-quarters share of coal in its primary energy supply, ever accede to an international agreement heavily taxing carbon emissions? Of course, China would not be a lone dissenter: the economies of Russia, Ukraine, Poland or South Africa are highly coal-dependent, and in the longer run crude oil combustion, and hence exports, would also suffer, prostrating a number of OPEC economies. How eager would these countries be to negotiate their economic demise?

A different kind of uncertainty is presented by the virtual impossibility to decide what is the true value of key economic indicators used to assess the prospects of effective environmental management. This applies above all to that generally accepted yardstick for measuring aggregate wealth, the gross national product (GNP), the total value of a nation's annual output of goods and services enlarged by resident's income from abroad and reduced by the corresponding income from foreign-owned activities. Weaknesses of the measure have been debated for decades, but it endures: its simplicity is its main attraction. Of course, for international comparisons the values in national currencies must be converted to a common denominator, and the US dollar still fills that role.

Conversions are done with prevailing exchange rates, and because during the past two decades these rates have been fluctuating, this practice results in such dubious outcomes as the World Bank's listing of China's 1989 per capita GNP at US $350 (World Bank 1991), that is less in constant monies than in 1981, at the beginning of Deng's economic reforms which brought unprecedented economic expansion! Using purchasing power parities (PPP), which reflect the actual buying potential, is a much more realistic choice than relying on simple exchange rate conversions – but PPPs, too, are misleading. A closer look at food consumption, a category which must be an essential part of any PPP comparison, illustrates the impossibility of doing it right.

If we just itemize the total food expenditures of a Chinese family and then find out how much this food would cost in the USA (or vice versa), we will

165

be doing something obviously unrealistic: average Chinese consumption of total food energy, as well as the intake of proteins, carbohydrates and lipids, is appreciably lower than in the US, and its composition is very different. Most notably, Chinese eat less than 20 kg of meat and less than 5 kg of sugar a year per capita, vs nearly 100 kg and 60 kg for Americans. To be equitable we should compare only the prices of identical foodstuffs – but the cultural differences affect markets in such profound ways that even these comparisons are unfair. For many basic foodstuffs there is simply no common basis, and others are available only as oddities. Similar differences arise in comparing the prices of household durables or common services.

As a result, various accounting and conversion methods put China's 1990 per capita GNP anywhere between less than US $400 and more than US $1600 (Smil 1993). If an international agreement would try to assign CO_2 control quota on the basis of national wealth, the Chinese would naturally favour the straight exchange rate conversion which makes them appear much poorer than they really are – while rich countries would favour the use of PPP adjustments which make poor countries richer. Another undecidable challenge is presented by the real rate of national saving, an indicator seen as an excellent marker of a country's commitment to its own future well-being. Americans score poorly, with the conventional measure of net national savings sinking below 5 per cent of GNP by the late 1980s. But alternative adjustments prepared by Eisner (1991) show that the real saving rates may be at least two and a half times and as much as more than five times higher than the standard account.

Finally, a basic intractable challenge influencing all of our attempts at valuing future benefits and losses, the matter of discounting. Traditional economic thinking was not concerned about intergenerational transfers: it assumed that increases in per capita productivity, resulting from greater human inventiveness and more efficient use of capital, would take care of coming generations. But once the economic process is seen in the perspective of environmental economics – merely as an open subsystem of a basically closed (except, naturally, for the input of solar energy) finite system of the biosphere – the question of obligations to future generations becomes morally unavoidable (Partridge 1981).

And so then is Solow's (1974) conclusion: 'The intergenerational distribution of income or welfare depends on the provision that each generation makes for its successors. The choice of social discount rate is, in effect, a policy decision about that intergenerational distribution'. But this choice can always be contested and exposed as not only inappropriate but positively harmful: there are simply no objective ways of discounting future values. Our limited life-span, as individuals and as communities, will perhaps always value the benefits accruing today or tomorrow higher than those years and generations ahead: as the benefit horizon recedes we apply higher discount rates.

But in order to manage the environment in sustainable ways the very opposite is needed. The benefits of preserving and replanting forests or cutting emissions of greenhouse gases are not only years but often several generations ahead, and only sufficiently low discount rates could guarantee today's rational investment. But very low discount rates can also encourage excessive investment and be socially counterproductive. In practice these dilemmas are solved politically, by moulding or gauging public preferences in democracies, and by imposing decision in totalitarian regimes. Obviously, this rather arbitrary *de facto* discounting does little to put long-term environmental management on a rational basis.

Finite amounts of non-renewable resources and the limited capacities of the environment to perform critical services mean that intergenerational concerns are largely a matter of zero-sum adjustments. But accepting this proposition would lead us to draconian conservation measures designed to maximize the duration of environmental resources. In practice, logic would then dictate the fastest possible transition to a truly sustainable society, one with a greatly reduced global population living strictly on the solar income and not tapping any accumulated natural capital. But we were there already as a species – as foraging bands or as shifting farmers – and our inordinately high encephalization drove us in environmentally more destructive directions.

Would we be willing to settle for good into this old-new non-expansive existence, even though it would be made much more acceptable because the new sustainable society would benefit from accumulated knowledge and would provide its members with a much higher level of material comfort? Many more questions can be asked about the consequences of this population and consumptive devolution, but the conclusion is clear: valuing the posterity is to enter a realm of ignorance and deep uncertainties – but concerns about sustainability are meaningless without at least implicitly putting a price on the future. And by the time we get there we will have to cope with many things we could not foresee.

Unforeseeable outcomes

As in all instances of complex interactions, there is a great deal of intrinsic uncertainties in predicting long-term consequences of management decisions regarding the biospheric change. And the surprises continue to happen even after such actions have been carefully considered and turned into much extolled policies. Excessive trust in reigning scientific consensus (as if majorities could be never wrong) or in elaborate computer models (as if machines able to handle vast amounts of data speedily would be inherently superior to critical reasoning) can easily result in regrettable developments.

Even a combination of solid scientific understanding, firm social commitment and tested technical ability is no guarantee of satisfactory outcomes as anticipated success brings very often unanticipated troubles and as promising solutions turn out to be merely beginnings of new problems. Simply, things can get worse when they get better. Alternative phrasings capturing this counterintuitive contradiction may be that advances beget regress or that, as Kenneth Boulding once put it, nothing fails like success. A perfect environmental policy example is the genesis of acid deposition over eastern North America and parts of Europe, a little appreciated case of a costly unforeseen change.

Well-publicized accounts of this environmental degradation start with sulphur and nitrogen oxides emitted from the combustion of fossil fuels, above all from large coal-fired power plants; carried with prevailing winds the oxides are eventually transformed into sulphates and nitrates which are deposited hundreds of kilometres downwind and may acidify lake waters and soils in susceptible regions and damage some vulnerable forests. A predictable response is to curb the emissions, especially those of sulphur dioxide, by installing expensive flue gas desulphurization units, and both the Western European nations and the United States and Canada are now committed to long-term programmes of large-scale emission reductions. Consequently, it is ironic to realize that acidification is largely a direct consequence of making things 'better' in the 1950s and 1960s.

In order to eliminate objectionable fly ash, highly efficient electrostatic precipitators were gradually installed on all large combustion sources, and tall stacks were built to disperse the remaining particulate matter and sulphur dioxide into large volumes of air in order to minimize ground concentrations of pollutants. These efforts paid off: in the United States total fly ash emissions in 1975 were only about two-fifths of the 1955 total and although sulphur dioxide emissions rose by about 30 per cent, ground concentrations of the gas declined almost everywhere owing to greater dilution. Clearly, air pollution from coal combustion got better and the air quality improved substantially.

Except, of course, for more acid precipitation in regions far downwind from large concentrations of coal-fired power plants! Why did this situation arise rather suddenly in the late 1960s and the early 1970s when very large quantities of coal had been burned ever since the mid-nineteenth century? Before strict particulate matter controls were put in place, alkaline elements in fly ash (calcium, magnesium, sodium and potassium forming between 15 and 45 per cent of fly ash mass) neutralized a large portion of acid sulphate anions (Smil 1985b). Alkaline cations are now largely captured in the removed fly ash, most acid anions are not neutralized and must be balanced by hydrogen cations to satisfy electroneutrality.

Concentration of hydrogen ions in precipitation increases, and the higher acidity damages sensitive aquatic biota and some forests far downwind because sulphur dioxide emitted in hot plumes from tall stacks is transported much farther than before and more of it can be oxidized to sulphate during that transfer. A laudable quest for cleaner air became a generation later the single largest cause of what is now seen as an important regional environmental degradation problem in eastern North America and in most of Europe. This counterintuitive twist is costing us dearly in new controls. But may we not discover that the sulphur dioxide clean-up now under way has brought a new, wholly unanticipated, environmental problem in yet another round of things getting worse as they get better? Indeed, there are already indications that the removal of SO_2 could become a major contributor to higher tropospheric temperatures.

Sulphate particles are among the most effective condensation nuclei in the troposphere, and as the total mass of anthropogenic sulphur compounds have increased at least sixfold since the year 1900, increased cloud cover reflects a higher share of incoming radiation and cools the troposphere. Wigley (1989) suggested that about half of the planetary warming attributable to increased concentration of greenhouse gases might have been offset by increased cloud cover generated by more abundant condensation nuclei above the Northern hemisphere. Charlson *et al.* (1992) make an even stronger claim. They estimate the current globally averaged climate forcing due to anthropogenic sulphate at about –1 to –2 W/m^2, comparable in magnitude to greenhouse gas forcing in the opposite direction, and hence largely offsetting any warming trend.

And analyses of long-term observations from more than 700 stations in the US, former USSR and China concluded that the Northern hemisphere's temperatures have increased during this century mainly at night (Karl *et al.* 1991). By far the best physical explanation for this unmistakable night-time warming signal is the sunlight-scattering effect of sulphate aerosols. This contention is also supported by the fact that the day-night temperature differences are much lower in Australia: only a small share of all anthropogenic sulphates is generated in the Southern hemisphere, and hence any warming should be more evenly distributed.

Another outstanding example of a counterproductive response is the mistaken identification of serious health risks from exposure to asbestos fibres and their subsequent costly removal from thousands of North American buildings. According to the US EPA (Environmental Protection Agency) estimates, asbestos removal from some 733,000 public and commercial buildings will cost at least US $(1988)53 billion (and perhaps as much US $150 billion) – while the best risk assessment now indicates that chrysotile asbestos, the dominant fibre in insulation, does not pose a health risk in non-occupational environments (Mossman *et al.* 1990). Exaggerated fears of PCBs (polychlorinated biphenyls) belong to the same category of miscalculations (Abelson 1991).

For perhaps the best example from the poor world I will point out the frustrations of dealing with widespread malnutrition in African, Asian and Latin American countries. This condition would seem to have obvious roots in shortages of protein or total dietary energy or both. Indeed, the syndrome became known as protein-energy malnutrition but careful studies have shown that protein deficiency rarely explains the condition (after all, human milk has a fairly low protein content, just 7.5 per cent, yet breast-fed babies thrive), that malnourished children are not ravenously hungry and do not eat every proffered food – and that even diseases, mainly recurrent infections, are not the key factor (Bhattacharya 1986).

Proper ways of infant feeding after weaning are in most cases more helpful in reducing the extent of protein-energy malnutrition than a seemingly common-sense effort aimed just at higher food intake. This conclusion is, of course, of great importance for estimating future nutritional needs – and hence environmental impacts – in poor, densely populated nations: maternal education, rather than simplistic expansion of food intakes, can become a superior, and clearly resource-sparing, means of better infant health.

And we will also be surely misled again and again because the real long-term concerns are also frequently unforeseeable. Preoccupations of the day seem to be so permanent yet a decade, or even less, can make them obsolete – but not before we will have committed considerable resources in dealing with what will be only marginal worries. As already noted in the first chapter, during the late 1960s America's leading environmental concern was the heavy nitrogen fertilization – but today's higher applications are hardly noticed. Shortly afterwards high mercury levels in tuna and swordfish led to fears about hazardously high contamination of marine food webs – but these concentrations were soon shown to be of overwhelmingly natural origin, caused by the predation and longevity of the affected species (Peakall and Lovett 1972).

Then the growing tanker traffic seemed to be putting oil blobs everywhere and oil spills were feared to damage coastal regions irreparably – but a survey of the global ocean found it to be in healthier state than a decade earlier and detailed studies demonstrated oil spills have little effect in the water column as evaporation, emulsification, settling and, above all, microbial oxidation keep the sea surprisingly clean (UNEP 1982). In contrast, there have also been major underestimates of environmental impacts, none of them more prominent than the already detailed chloro-fluorocarbon (CFC)-driven destruction of stratospheric ozone. Naturally, such shifting conclusions and worrying surprises will make it even more difficult to find consensus for effective preventive or remedial actions.

LIMITS OF ACTION

Well, in our country', said Alice, still panting a little, you'd generally get to somewhere else – if you ran very fast for a long time as we've been doing!' 'A slow sort of country!' said the Queen. 'Now, here, you see, it takes all the running you can do, to keep in the same place'.

Lewis Carroll, *Through the Looking Glass* (1870)

The Red Queen's explanation may be correct more often than we think also on this side of the looking-glass. And not infrequently things may be even worse than that as we actually fall further behind in spite of some rather impressive running. Obviously, there is an enormous range of reasons for these limits to effective action, including both a large number of universal factors as well as a multitude of site- and time-specific causes primarily determined by environmental variability and socio-economic arrangements.

These impediments will be extremely important in determining eventual rates and degrees of effective change. In this assessment of limits I will once again concentrate on key existential realities and look at the most important obstacles and complications weakening or blocking the possibilities of successful responses. I will subdivide this appraisal into four categories dealing with demographic imperatives, with environmental givens and feedbacks, with the inadequacies of technical fixes, and with the challenges of socio-economic adjustment. In reality, most limits are clearly multi-factorial, and dynamic relationships among population, environmental, technical and socio-economic factors often make an easing or elimination of a particular obstacle, conceivable on its own, an unlikely or even an impossible choice.

Imperatives of population growth

Essentials of the grand planetary split are well known, but, I always feel that they are not truly appreciated. Numbers get cited, projections are updated, data sets are computerized or put on CD ROMs (compact disc read-only memory) – but none of this can convey the crush of bodies on Shanghai streets or Egypt's disappearing and salinizing deltaic soils. Even walking Shanghai's overflowing pavements or the muddy, manure-spattered alleys of Egyptian villages is not enough. Only living there an unpriviliged life of a permanent exile from the rich world's affluence would furnish the proper perspective.

But the numbers are telling if thought about, and not merely cited. Of the 1990 world-wide total of 5.3 billion people only about 1.1 billion lived in industrialized countries: 375 million in affluent Western Europe, about 275 million in, on the average, even more affluent North America, 125 million in Japan, 20 million in Australia and New Zealand; the rest was in the comparatively poorer nations of Central and Eastern Europe (125 million), and in Russia and the European part of the former USSR (about 220 million). Of the almost 4.2 billion people living in poor countries more than two-thirds were in low-income economies where even basic existential needs could not be guaranteed for large shares of population (World Bank 1991).

Every attempt to maintain the integrity of the Earth's environment must be mindful of this basic division – and of the further inexorable skewing of that already uneven distribution. Long-term population forecasts are usually

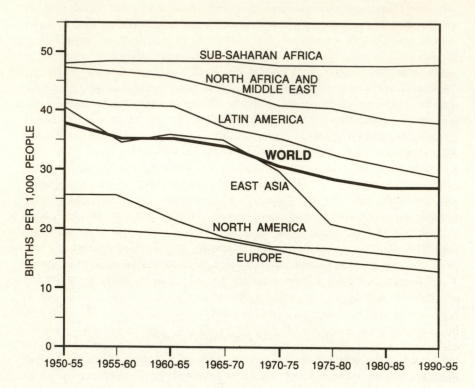

Figure 5.1   Fertilities have declined everywhere since the middle of the 20th century
except in sub-Saharan Africa.
Sources: Based on Keyfitz (1989) and World Resources Institute (1992).

inaccurate, but looking just a generation, or only a bit more, ahead carries
tolerable errors because there can be few fundamental shifts in the
underlying dynamics of population growth. The natural rate of increase has
been declining steadily in Latin America and in most Asian countries, but
because it applies to an expanding base the absolute annual increments are
still growing. This is true also in China in spite of an unprecedented post-
1972 drop in birth-rates. Even when they decline substantially, total fertility
rates in many Asian rural areas have a tendency to get stuck well above two,
largely as a result of the traditional desire to have at least one son (Arnold
and Liu 1986; Chowdhury and Bairagi 1990).

The only continent where high fertility shows few signs of decline is Africa
(Figure 5.1). This persistence stems largely from a traditional pronatalist
belief system identifying continued fertility with virtue and associating
reproductive failure or restriction with punishment and evil (Caldwell and
Caldwell 1987). There seem to be little chance that any substantial shift will
occur during the 1990s, and the population of sub-Saharan Africa may

actually surpass both the World Bank and UN projections for the year 2000. The only highly uncertain factor which may lower the total appreciably is the continuing diffusion of AIDS, but the already high incidence of this disease in central and eastern Africa is in itself an enormous and intractable environmental catastrophe (Anderson and May 1992; Weeks 1992).

Aggressive family planning programmes could slow down the future increases, but during the 1980s no African country had a strong commitment to population controls (Keyfitz 1989). Weak to moderate efforts were under way only in a number of Arab African countries (Egypt, Tunisia, Algeria, Morocco) and also in Kenya, Tanzania, Liberia and Senegal. There was practically no effort to control population in more than 20 sub-Saharan countries. And, as shown in the Nepali case (Metz 1991), the existence of a population control programmme may have virtually no impact when local microeconomic conditions continue favouring large families.

The negative environmental consequences of continuing high absolute (and in Africa also high relative) rates of population growth throughout the poor world can be denied only by providers of techno-optimistic delusions. These impacts will be even more damaging because most of the growth will come in already badly degraded ecosystems, and in regions affected by recurrent droughts. Nearly half of the world-wide population increase (3.2 billion people) forecast between 1990 and 2025 will come in just six countries: India (almost 600 million), China (350–400 million), Nigeria (150–180 million), Pakistan (close to 150 million) and Bangladesh (100–120 million).

There is no need for complex forecasting exercises to see that even some extraordinarily successful efforts promoting conservation and economic efficiency would be swamped by these increases, and that it will not be possible even just to stabilize the demand for energy, water, minerals and wood. The double stricture of huge and growing populations and of an already badly degraded and steadily deteriorating environment makes the chances of environmental recovery during the next generation very unlikely. The best realistic expectations would be keeping the rates of environmental degradation not much above the current levels.

If a poor populous country will more than double its population before the year 2025, its fertilizer applications and, if in arid climate, its use of irrigation water – and hence the attendant environmental impacts of siltation, salinization and pollution – will have to increase even if the average water and fertilizer use efficiency would go up by unlikely margins of 50–80 per cent. While American or Russian agronomic malpractices could be set right with a large-scale adoption of environmentally benign techniques without reducing the average availability of food, farming goals in the six large poor countries will be dominated by the quest for higher yields, an effort inconceivable without higher inputs and more extensive environmental degradation.

173

Similarly, unlikely energy efficiency gains of between 30 and 50 per cent would still require larger fuel and electricity supply in most poor populous countries even if their very low per capita use would barely increase for the next three decades. Because a large part of the population growth will have to be absorbed by expanding cities, greater concentrations of environmental and social decay will present often insurmountable problems to local or national management capabilities. In 1950 eight out the world's ten largest cities were in industrialized countries. In contrast, by the year 2000 only two among the world's ten largest metropolitan areas – Tokyo and New York – will be in rich countries. Three will be in Latin America (Ciudad Mexico, Sao Paulo and Rio de Janeiro) and five in Asia (Shanghai, Beijing, Bombay, Calcutta and Jakarta).

A fundamentally different outlook for managing future environmental stresses in rich and poor nations – and a number of economic and political implications arising from this disparity – cannot be illustrated better than by looking at US and Chinese contributions to the global generation of greenhouse gases. The combination of America's extraordinary affluence (in terms of purchasing power parities the nation is still well ahead of Japan and Western Europe) and its wasteful use of materials and energy (among the rich countries only Canadians need more energy to produce a unit of GDP) could make it possible to cut the emissions of greenhouse gases by up to 30–40 per cent with little or no long-term economic cost (Committee on Science, Engineering, and Public Policy 1991) – and without any destabilizing drop in the average quality of life.

China has also considerable inefficiencies in energy and material utilization as well as in its agronomic practices, but existential imperatives, capital shortages and resource availability will greatly limit the nation's scope of remedial action for at least the next two generations (Smil 1993). But even if such control efforts were to be successful, their effects would be easily negated by continuing population growth and by rising personal consumption. In China's case the combination of these realities will make it impossible for the next two to three decades even just to stabilize the total emissions of greenhouse gases.

Simple estimates are sufficient to outline the most likely outcomes. The USA will not have large per capita increases of primary energy use during the next generation. The country's total CO_2 emissions in 1990 were less than 1 per cent above the 1980 level, and in per capita generation actually declined by about 10 per cent (Boden et al. 1990). Assumption of an additional 20 per cent per capita decrease during the coming 30 years is hardly excessive. The US has no firm long-term reduction goals, but a number of western countries aim at 20–25 per cent cuts from the late 1980's levels by as early as the year 2000 and no later than 2025 (Morrisette and Plantinga 1991). Coupled with the medium variant of UN population projection (UN 1991), a 20 per cent cut would translate to total annual US emissions of about 4.5 Gt (billion tonnes) CO_2 in the year 2020.

In contrast, China's total CO_2 output rose by nearly 50 per cent during the 1980s, a 30 per cent increase in per capita terms. Similar increases must be expected during the 1990s, and even if this growth moderates after the year 2000, and if China's per capita consumption of fossil fuels would be no more than 40 GJ/year in the year 2020 (60 per cent above the 1990 rate, but still 30 per cent below the world mean of 1990), the country's total CO_2 generation would be at least 5.5 Gt CO_2, 20 per cent above the US total. Particulars will be inevitably different but it is virtually certain that China will at least rival, and very likely it will surpass, the US CO_2 generation to become the world's largest producer of the gas by the year 2020.

The combination of demographic and developmental imperatives makes it also certain that the poor world's emissions of CH_4 and N_2O will also be rising. Again, the Chinese example is revealing. Emissions of the two greenhouse gases are the result of an extraordinary challenge of feeding more than one-fifth of the global population from less than one-fifteenth of the world's arable land. China's average per capita food availability is now above the minimum needs, but large provincial supply disparities and major harvest fluctuations caused by droughts and floods mean that malnutrition remains a major concern. In addition, there is a large unmet demand for animal foods. In order to produce 1 kg of meat the Chinese require about 4 kg of grain (Tuan 1987) and the country thus needs substantial increases of both food *and* feed grain production merely to eliminate the present deficiencies. And even larger increments will be needed to accommodate huge population increases.

Should the average intakes stay at the 1990 level, feeding an additional 150 million people by the year 2000 will require an incremental food supply equivalent to the total current food consumption in Brazil, and taking care of an additional 300–350 million people a generation later would call for a supply expansion roughly equivalent to the 1990 combined harvests in Indonesia, Pakistan and the Philippines (FAO 1990). Moreover, the greatly expanded harvests will have to come from a smaller area of arable land whose quality will be declining: the only way to produce higher harvests is further intensification of China's already highly intensive cropping. China is already the world's leading user of nitrogenous fertilizers and the largest cultivator of rice, and it is almost certain that during the next two generations the country's share of N_2O and CH_4 will rise above the shares (between 20 and 30 per cent) of the late 1980s.

Rich countries can step in and finance the replacement of China's growing CFC production – but there is little the west can do about the inevitable increase of China's fertilization, rice-growing and coal combustion. The combination of large absolute population increases, and the necessities of feeding huge populations and supplying larger quantities of fossil fuels and electricity in order to energize the much needed modernization will lead inexorably to higher emissions of CO_2, CH_4 and N_2O even if the average improvements in the quality of everyday living would be rising very slowly. By the year 2020 China may account for one-fifth of the global generation of greenhouse gases, double its current share.

Demographic imperatives combined with the quest for a higher standard of living will bring analogical trends for other forms of environmental degradation. During the coming generation the rich world can become much more efficient: by either capping or even substantially reducing its pollutant releases and its ecosystemic changes it will be responsible for a steadily declining share of undesirable world-wide environmental change. In contrast, poor countries will, almost without exception and even if doing their best to become less wasteful and more considerate, move in the opposite direction. Only an eventual stabilization of their populations and the provision of a decent quality of life could moderate, and later reverse, this trend.

But that time is still beyond the rational planning horizon. And so for generations to come rich countries will continue facing the pointed moral and economic dilemmas of how to deal with the increasing migration propensities of poor world populations, a trend which may be locally and regionally aggravated by the creation of more environmental refugees fleeing badly degraded and overpopulated places. Hardin (1974) argued strongly that as long as there is no true world government controlling reproduction everywhere, it is impossible to survive with dignity if the Spaceship Earth ethics would be our guide (that is, take everybody on board). He had no doubts that 'for the foreseeable future survival demands that we govern our actions by the ethics of the lifeboat'.

No matter which strategy a rich country will pursue, rising numbers of migrants seeking a better life will be straining its socio-economic foundations. Today's examples, ranging from Mexican influx to the USA, Arab migrations from North Africa to Europe and Bangladeshi overflow into Assam, West Bengal and beyond, may look modest in their totals in 20–30 years. Not surprisingly, hard questions of long-term population growth are not preferred topics in international discussions on global environmental change. But the euphemistic circling of continuing high population growth in poor countries should not be accepted in any responsible discussion of global environmental problems: it is self-evident that such countries – Haiti or Nigeria, Bangladesh or Laos – would greatly benefit from reduced fertilities, not only in economic and environmental terms, but also in terms of basic human dignity.

Pointing out, correctly, the relatively minor role of poor countries in global environmental degradation – the resource claim of an additional child in a rich nation being an order of magnitude above that born in a poor country – is no rational argument for continuation of high fertilities inevitably reflected in further degradation of natural resources and in higher pollution releases. But this approach makes it virtually impossible not to shift international discussions of population growth almost immediately into the questions of economic development. This is both desirable and unfortunate – and this unsatisfactory reality is not going to change soon.

Environmental constraints and feedbacks

Inescapable as the reality may be, many would-be managers of the global commons keep forgetting how limited our actions are by fundamental environmental constraints as well by many inexorable ecosystemic feedbacks. Even the prodigious radiation flux which energizes the biosphere has its limiting aspects. Solar radiation has gradually increased with the ageing of our star, but on the civilizational time-scale the input of energy – except for some minor short-term fluctuations (Frohlich 1987) – is constant, averaging nearly 1370 W/m^2. This flux is too high to be a limiting factor in photosynthesis, but if we were to supplant fossil fuels with renewable energies, highly unequal spatial distribution of direct solar radiation, and even more unequal power densities of indirect solar flows, would greatly restrict our energy options.

Several regions which would be among the greatest beneficiaries of converting direct solar radiation have surprisingly limited potential. Sichuan Basin, one of the world's most densely settled agricultural plains supporting most of the province's 110 million people, has 220–50 cloudy days a year, dense fogs come, on the average, every third day, it rains on nearly 150 days, and winters average about 1.5 hours of sunshine a day (Domrös and Peng 1988). Prospects of boiling water for the breakfast tea and cooking rice gruel with solar devices are, literally and figuratively, hardly bright. Similar solar deprivation is a norm in large parts of other poor populous nations, notably in Brazil and Nigeria (Figure 5.2).

Hydroenergy capacities are distributed very unevenly: three countries – China, Russia and Brazil – have about two-fifths of exploitable potential (World Energy Conference 1986). Even relatively windy places have long spells of calm and many densely populated areas are not windy at all. Throughout large parts of South-east Asia windmills may be idle for ten months a year, and a survey of India has not found any location with average annual wind speeds in excess of 18 km/h while in the mid-latitudes, where most of the rich countries are situated, such windspeeds are very common (Merriam 1977). Maximum photosynthetic efficiencies are reduced by inevitable respiration losses to about 5 per cent for plants in ideal environments, by common shortages of water and nutrients to only about 1 per cent in temperate and tropical rain forests, and to an average of just 0.3 per cent for all continental ecosystems.

As yet we cannot increase photosynthetic efficiency. We can just manipulate the process to harvest a higher yield by growing plants with inherently

177

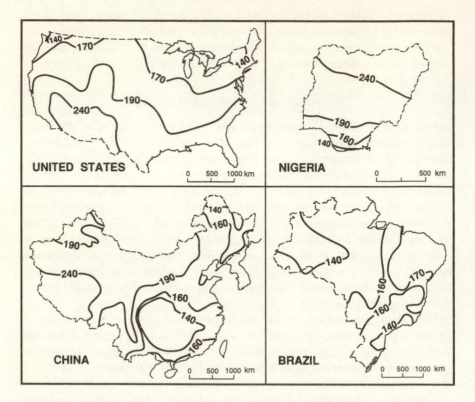

Figure 5.2   Solar deprivation of tropical, and also some monsoonal subtropical, regions is illustrated by comparing average insolation (all values are annual means in watts per square metre) for the USA, China, Brazil and Nigeria.
Sources: Compiled from a variety of climatological sources and converted to watts per square metre.

lower respiration losses (such as corn or sugar-cane), by breeding which redistributes the photosynthate to harvested parts, and by providing an array of subsidies (irrigation, fertilization, pesticides) which maximize the harvestable storage. Generally only food crops receive such costly inputs, and our capacities to increase non-agricultural primary production remain very limited.

This fact eliminates any realistic possibilities of vastly expanding photosynthetic capabilities of the biosphere. For example, the Norwegian Conservative Party proposed to use a small part (1–3 per cent) of gross oil earnings to expand forests which would eventually sequester the CO_2 produced by the combustion of oil. This proposal, advocated in different versions by many other green enthusiasts, has the obvious advantage of a long-term ecosystemic and economic gain. But to sequester just half of all carbon released by the combustion of fossil fuels – some three Gt a year –

178

would require (with a far from conservative annual productivity mean of 5 tonnes (t) of phytomass, or about 2.2 tonnes carbon/ha) additional 1.4 Gha (billion hectares) of trees, an area equivalent to half of the world's closed forests!

There is a profusion of other examples of environmental constants, or rather narrow rates, which limit our quest for higher productivities or hinder our efforts at better environmental management, from recalcitrant legumes to touchy bacteria. Legumes may supply their own nitrogen and are an excellent source of food protein – but one of the most important biochemical constants is their low average productivity. As a result, the average world-wide yield of all legumes went up by only about 40 per cent in four decades (FAO 1950–91). In contrast, average world-wide cereal yields have increased during that period nearly 2.2 times. But fertilizing legumes will not do: their symbiotic nitrogen fixation declines with available nitrate in the soil (Stevenson 1986).

And it would be a great multipurpose boon for many parts of the fuel- and food-short poor world if animal and human wastes and some crop residues could be anaerobically fermented yielding clean biogas for cooking and high-quality pathogen-free organic fertilizers for the fields. But methanogenic bacteria will work well only in strict anaerobicity, at temperatures around 30°C, with near neutral pH, and with C/N ratio around 30 (Smil 1983). Departures from those optima – a small crack in a clay digester, cold nights or seasons, low pH, too much straw in the fermenting mixture – cut the biogas yield or stop the fermentation entirely. The Chinese, who are pioneering the mass diffusion of biogas production, learned these lessons the hard way. In the late 1970s, when they built some 7 million digesters, they had plans for 75 million units for the mid-1980s – but by 1990 they had no more than 5 million of them in operation (Qiu *et al.* 1990).

There is little mystery about most environmental constants and performance ranges: it is rather their ubiquity which limits our management options. In contrast, a solid grasp of ecosystemic dynamics remains elusive. If to manage ecosystems successfully is to maintain their stability, then we should be able to assess their resistance to disturbances and their capacity for adjustment. This could not be done without identifying first their stable equilibria – and yet an extensive analysis shows that there is no clear demarcation between natural assemblages that may exist in an equilibrium state and those that do not (Connell and Sousa 1983). We also have only a fragmentary understanding of internal interactions which in many ecosystems generate strong positive feedbacks and largely determine their stability and productivity.

For example, only during the 1980s did we come to realize the criticality of photosynthate transfer to plant roots where it is used by mycorrhizal symbionts or released into the surrounding soil where it supports complex communities of fungi and micro-organisms. Of nearly 7000 species of

179

studied angiosperms, 70 per cent are consistently mycorrhizal, including all coniferous trees. Not surprisingly, these strong plant-soil links can be turned into incapacitating weakness under stress.

An outstanding example is the contrast between forested and clear-cut areas in high elevations of the Klamath Mountains of Oregon and California (Perry *et al.* 1989). Standing forest, dominated by old white-fir growth is classed in the highest productivity grade for its elevation, but adjacent areas, clear-cut during the 1960s, have not been reforested in spite of numerous attempts. Similar cases of irretrievably lost productivity after the disruption of strong plant-soil feedbacks have been described in former woodlands surrounding Lake Victoria and in the grasslands of western North America.

One of the latest additions to this category of destroyed feedbacks is the accelerated degradation of arid lands. Based on the studies in the Jornada Experimental Range in southern New Mexico, Schlesinger *et al.* (1990) identified a clear positive feedback in association with long-term grazing of semi-arid grasslands. This practice causes an increase in both spatial and temporal heterogeneity of soil resources (above all, water and nitrogen) which promotes invasions by desert shrubs. While soil resources are further concentrated under shrub canopies, erosion and gaseous emissions keep on degrading soil fertility from the barren areas between shrubs. Identical processes operate in other semi-arid and arid regions which now cover about one-third of all continental surfaces.

Strong interactions between plants and their rhizospheres and mycor-rhizospheres in tropical rain forests are one of the principal reasons complicating, or excluding, sustainable land management alternatives. Devising continuous cropping schemes for the humid tropics is a perfect example of this limitation. While it is obvious that the traditional extractive pioneering farming can be of only a very limited duration, a simple transfer of mid-latitude practices is not acceptable either (Eden 1990). Even soils with good structure and relatively high nutrient content will suffer from compaction, erosion and nutrient loss once the tree cover is removed and the surface is bared to heavy rains and to machinery.

In poorer soils it is very difficult to maintain adequate fertility. Experimental continuous cultivation of corn and rice (rotated with soy beans and peanuts) in Peru required annual inputs of 80–100 kg/ha of nitrogen and potassium, 25 kg/ha of phosphorus and magnesium as well as regular additions of copper, zinc and boron (Sanchez *et al.* 1982). Even if these nutrients could be transported in requisite amounts to remote locations, such fertilizing rates are far beyond the capital means of tropical subsistence farmers.

Another fundamental limitation to cropping in the humid tropics is the inherently higher presence and more vigorous activity of pests. High temperatures and high humidities promote continuous and rapid growth and spread of viruses, bacteria, fungi and insects, and this heterotrophic attack is naturally intensified by planting extensive monocultures. And because monocultural tree crops appear to be as vulnerable as annual plants, opportunities for silviculture are also limited. Clearly, a combination of natural constraints makes it exceedingly difficult to devise an intensive-cultivation agro-ecosystem thriving in the humid tropics.

And we do not really know how far we have gone in destruction and degradation of the biosphere. Many communities have been obliterated by settlements, fields, erosion and desertification, and more than a third of the continental surface has been altered to a noticeable degree by human actions. But even a detailed inventory of these changes would tell us little (with the obvious exception of eradicated endemic formations or species) about the irretrievable losses and about the long-term capacity for restitution. But this knowledge would be essential for optimized management of the remaining riches.

What we do know is hardly encouraging – and these realities also limit our management options. In many ecosystems destruction and disruption has been already so great that even a strict protection of all remaining natural areas would not be sufficient to preserve many species. There is a general agreement among the students of conservation biology that most existing national parks and reserves are far too small to guarantee long-term survival of especially large species (Simberloff 1988). Indeed, Newmark (1987) claims that even North America's largest national parks (Wood Buffalo in Alberta has 44,807 km^2, but Yellowstone only 8,983 km^2; for comparison, Serengeti in Tanzania 12,950 km^2) are too small and as a result have already lost about one-quarter of their large mammals. And because North America has still the largest areas of wilderness, the situation is considerably worse on all other inhabited continents.

An inventory of the amount of wilderness remaining in the world (with wilderness defined as undeveloped land in blocks larger than 400,000 ha) shows that about one-third of the continental surface is still shaped almost exclusively by biospheric processes – but most of this area is in highly stressed, low-productivity ecosystems (Figure 5.3). About 40 per cent of the total are in circumpolar tundra communities, and another 20 per cent are in subtropical deserts and semi-deserts, while only about 6 per cent of the remaining wilderness during the mid-1980s was in tropical rain forests, and less than 2 per cent in temperate (rain and broadleaf) forests (McCloskey and Spalding 1989). Moreover, only a small fraction of these wilderness areas (generally less than 20 per cent in every major biome) are currently protected.

Ecosystemic realities also limit the opportunities for an effective compromise between preservation and utilization of species-rich ecosystems. In order to preserve large contiguous areas of remaining tropical moist forests some researchers extol the benefits of their sustainable use by means of traditional collecting of nuts and medicinal plants, rubber tapping, hunting

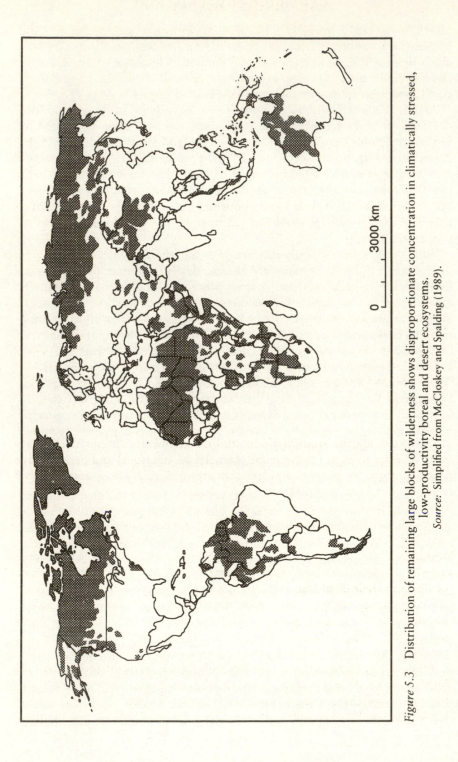

0
3000 km

Figure 5.3   Distribution of remaining large blocks of wilderness shows disproportionate concentration in climatically stressed, low-productivity boreal and desert ecosystems.
Source: Simplified from McCloskey and Spalding (1989).

and fishing (Fearnside 1989). Selective tree-felling with careful removal of merchantable boles which would leave roots, stump, bark and branches on the site to help the regeneration of plant cover would be another preservation option.

In 1985 Brazil's National Council of Rubber Tappers proposed the creation of extractive reserves in Amazonia, and by 1990 the Brazilian government had plans to set aside 25.7 Mha of the Amazonia for *extrativismo*, that is, mainly for removal of such non-wood forest products as latex, resins and nuts. This is certainly an environmentally appealing management option – but ecosystemic imperatives make it also an inherently low-yield one, and hence of marginal financial and commercial attraction (Goodland *et al.* 1991). Sustainable logging would require exploitation of much larger areas, and tree removal along longer logging trails is not only more costly but it can lead, even when done as carefully as possible, to considerable local vegetation damage and erosion.

Naturally, some of the environmental limits to effective management will be modified in the future by our inventiveness. We may eventually have high-yielding legumes, or nitrogen-fixing cereals, or hardy methanogenic bacteria: promise of bioengineering has been too often exaggerated – but it should not be underestimated either, especially not the long-term one. But even if we would have such successes, they will surely coexist with at least as large a number of technical failures – or innovative successes which later turn out to be just beginnings of new problems.

Inadequacies of technical transformations

To demonstrate the restricted potential of technical solutions there is no need to refer to such fundamental obstacles as structural and performance limits imposed by material properties, thermodynamic laws, probabilities of component failure and maximum capacities and densities (van Wyk 1985). Less fundamental, but very frequent, limitations will make it difficult to translate many of today's promising technical fixes into tomorrow's widely adopted realities. Contrary to popular impressions created by techno-enthusiasts and shared by many environmental activists eager to promote 'green' solutions, the costs of many desirable innovations will be a major obstacle to their diffusion. Obviously, many techniques show the expected lowering of unit costs with continued development and spreading adoption, but in many cases such reductions still leave the price too high or can make only a marginal difference because of the severity of the underlying problem.

Efficiency gains in irrigation offer perfect examples of often prohibitively high costs of desirable innovation. In the USA typical costs of install-ing centre pivots are close to US $(1990) 10,000/ha compared to just US $(1990) 400/ha for controlled flooding systems. Moreover, centre pivots are energy-intensive, both in terms of their construction and operation (Batty and Keller 1980). The latter cost will depend, obviously, on the depth of the tapped aquifer and the crop grown: growing corn with water drawn from the Ogallalla aquifer needs anywhere between 10–100 GJ (gigajoules)/ha (Smil *et al.* 1983). Israeli figures show that the energy cost of a simple portable aluminum sprinkler pro-rates annually to about 300 MJ/ha, but that solid-set drip lines amortize at 14 GJ/ha, nearly 50 times higher (Stanhill 1990b).

Substituting a traditional (55 per cent efficient) surface irrigation with a drip system (85 per cent efficient) involves a net energy investment of 39 MJ (megajoules)/m^3 of water, an equivalent of almost 1 kg of crude oil, and the cost is repayable only by growing such very high value crops as fruits or flowers. And as impressive as Israeli advances have been, they are insufficient to maintain a competitive production on a sustainable basis. Given the limited amount of water resources available, Stanhill (1990b) estimates that, in order to keep irrigating half of the cultivated area, water use efficiency would have to double by the year 2000. Such a gain could not be achieved without major progress toward decoupling transpiration and yield formation, a formidable and so far elusive task for plant breeders. Basic research looks at increasing plant resistance to the diffusion of water vapour or reducing its absorption of solar energy, but early applications are unlikely.

Interdependence of technical factors is another common obstacle to speedy and inexpensive innovation. A seemingly marginal change of a single variable in a complex system built over a long period of time with enormous capital investment may lead to profound alterations of the whole set-up. America's new Clean Air Act of 1990, designed to limit emissions of SO_2, will cause enormous problems in particulate control. Performance of electrostatic precipitators is notoriously sensitive to changes in heat rate and SO_2 content in the effluent gas and so the problems of adjusting the well-established fly ash controls will be perhaps more significant than those of newly imposed desulphurization (Rittenhouse 1990).

The introduction of substitutes for CFCs is another perfect example of such a complicating, destabilizing effect. Even for those applications still requiring fluorocarbon-based compounds there are many closely related hydrochlorofluorocarbons (HCFCs) and hydrofluorocarbons (HFCs). But selecting the most appropriate compounds, and introducing them to equipment which in the US alone is estimated to be worth US $(1990)135 billion and has an expected lifetime of 20–40 years will be a profound transformation. Problems of material compatibility, product-life cycles and energy efficiency will complicate a rapid adoption (Manzer 1990).

For example, the use of CFC-substitutes in building and appliance insulation could lead to a notable increase of US energy consumption – by as much as an equivalent of heating about 30 million homes for a year – if the already available alternatives (above all, $CHClF_2$) are used (Baxter and

Fairchild 1990). On the other hand, some emerging ozone-safe techniques would actually lower the total energy consumption – but they will not be available for mass diffusion as rapidly as the latest CFC-elimination agreements require. Limits of this technical fix are clear: delays may lead to greater environmental damage, but haste may cause additional waste, and hence also additional environmental degradation.

Limits to technical solutions are imposed even more often by economic and social arrangements. In many poor countries adoption of efficient irrigation will be limited less by high capital and operating costs than by the incompatibility with existing land tenure and cropping patterns. For example, the average size of fields in East Asia is well below 0.1 ha, an area two orders of magnitude smaller than the typical coverage by centre pivots, the most efficient method of grain irrigation. Socio-economic realities also impose two practical limits on energy conservation: diffuse ownership of energy convertors makes it impossible to achieve rapid savings in non-industrial consumption, and it also necessitates widespread participation before achieving substantial cumulative gains. Energy conservation has few equivalents of exciting giant oilfield discoveries, no analogies of dramatic announcements of technical breakthroughs promising an endless source of cheap power.

In poor countries where private energy consumption is a small fractional of total fuel and electricity flows, rationalization of industrial conversions can bring relatively rapid and impressive results. But in rich countries, where final energy uses are much more evenly divided among industries, transportation, services and households, effective conservation efforts require an extensive individual participation. Even such obviously major conservation measures as higher efficiency standards for cars, air-conditioners or refrigerators cannot – if undertaken alone – reduce national energy use by large margins. For example, a 20 per cent cut in US consumption of gasoline would have reduced the country's total 1990 primary energy use by only about 3.5 per cent (BP 1991).

Moreover, when pay-off periods for investment are relatively long, and when marginal returns cannot give a clear incentive to save, conservation's aggregate success is impossible without long-term commitment. Longer amortization spans justify higher energy inputs and generate greater life-cycle savings – but in modern economies this goes against the dominant preference for the least first cost, and in North America it goes also against the still high residential mobility. In more stable societies, where houses are commonly bought only after many years of saving and are frequently inherited, long-term conservation commitments are much more appealing.

A number of promising technical solutions are rejected by society because they were found to have either demonstrably higher risks than the techniques which they were to replace, or because there is a widespread fear of their potential impacts. Nuclear electricity generation is, of course, by far the best illustration of this phenomenon. After tens of billions of dollars in R&D subsidies, and after rapid commercialization starting in the late 1960s, it should have greatly benefited first from the OPEC-driven energy price increases and acid precipitation concerns of 1970s and then even more from the concerns about possible global warming.

Nuclear establishments still believe that fission is the best way to reduce CO_2 emissions (Pendergast 1992) – but who else shares that belief? The capital costs of new North American nuclear power plants are appalling, they take too long to build, and management of many stations has been incompetent. Public opposition, in the aftermath of Chernobyl, is high in every rich country, and no nation with sizeable nuclear generating capacity has solved the problem of long-term disposal of radioactive wastes. Consequently, nuclear generation is highly unlikely to recover even in a greenhouse (Ahearne 1989).

Combustion of municipal wastes is another excellent example of such contradictory realities. About two-thirds of typical city rubbish is made up of organic compounds, with heat contents of individual constituents ranging from just 4 MJ/kg for vegetable refuse to just over 44 MJ/kg for polyethylene. Average heat rate of the combustible portion is around 10 MJ/kg. This is roughly equivalent to two-thirds of dry wood's value, and to a third of heat content in standard coal. Besides providing electricity and steam (rubbish-fuelled co-generation would be a particularly attractive conservation option), waste combustion also greatly reduces demand for increasingly scarce landfill sites, while pre-combustion sorting can recover a variety of recyclable materials.

Why then is there no rush for rubbish combustion? The principal reason is the generation of potentially carcinogenic dioxins and furans, as well as arsenic, beryllium, mercury, nickel, cadmium and chromium. Releases of heavy metals may be as much as five orders of magnitude more hazardous than the emissions of organics (Steverson 1991). Proper ash disposal is another concern. Consequently, in spite of obvious environmental benefits and in spite of regulatory and industrial acceptance, waste incineration has not been widely accepted by the public, and its proposed use is a matter of emotional confrontations (Washburn *et al.* 1989). Perhaps the only exception is Japan, a country with an acute shortage of disposal sites (Herskowitz and Salerni 1989).

But certainly the most important case of technical interdependencies and socio-economic obstacles limiting a rapid introduction of environmentally desirable innovations is the profound difference in power densities of fossil-fuelled and solar energy systems. In terms of typical power densities a solar civilization based on a combination of direct radiation, hydro, wind and biomass energy conversions would be the opposite of current arrangements.

Critical rates illustrate these disparities. Typical power densities of existing techniques are between 20 and 60 W/m^2 for direct solar energy capture,

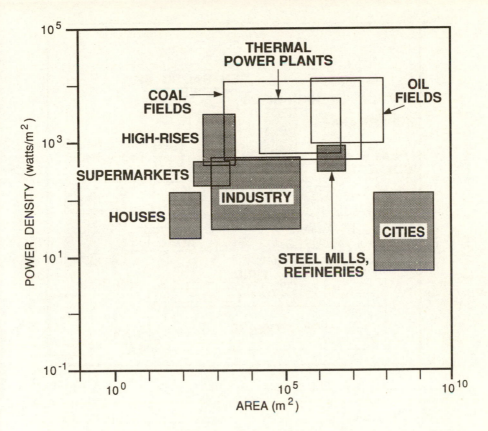

Figure 5.4 Power densities of fossil-fuel extraction and thermal power generation are orders of magnitude above the typical power densities of energy consumption in modern societies.
Source: Based on Smil (1991a).

10–50 W/m² for geothermal, tidal and upper-course hydrogeneration, just above 1 W/m² for most of the lower-course hydrogeneration requiring large reservoirs, and below 1 W/m² for biomass energies. In contrast, extraction of fossil fuels and thermal generation of electricity produce commercial energies with power densities orders of magnitude higher, ranging mostly between 1 and 10 kW/m². Final utilization power densities range mostly from between 20 and 100 W/m² for houses, low energy intensity manufacturing and offices, institutional buildings and urban areas. Supermarkets and office buildings use 200–400 W/m², steelmills and refineries 300–900 W/m², and high-rises up to 3 (kilowatts)/m².

Fossil-fuelled societies are thus diffusing concentrated energy flows: they have been producing fuels and thermal electricity with power densities one to three orders of magnitude higher than the common final use densities in

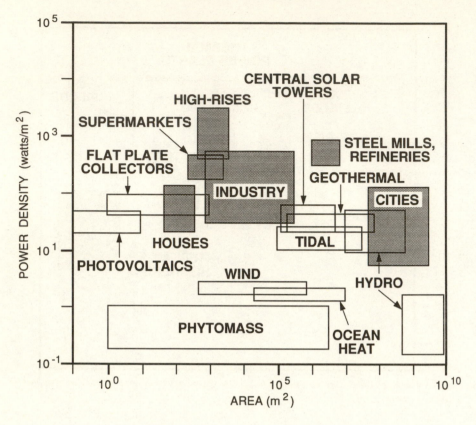

Figure 5.5 In contrast, power densities of direct and indirect solar conversions and of tidal and geothermal energy are orders of magnitude lower than the common consumption power densities, a reality which would necessitate a major restructuring of modern civilization.
Source: Based on Smil (1991a).

buildings, factories and cities (Figure 5.4). As a result, space taken up by extraction and conversion facilities is relatively small in comparison with transportation and transmission rights-of-way required to deliver fuels and electricity to consumers. A solar society inheriting the existing urban and industrial infrastructures would have to concentrate diffuse energy flows. Only for some final uses (most notably for house heating) would it harness its energies with the same power densities with which they would be used, but typically it would have to concentrate diffuse flows in order to bridge power density gaps of two to three orders of magnitude (Figure 5.5).

Mismatch between the low-power densities of renewable energy flows and the relatively high-power densities of our final energy uses means that any large-scale diffusion of solar energy conversions will require a profound spatial restructuring engendering major socio-economic transformations. This would not only vastly increase the fixed land requirements for primary conversions, but it would also necessitate more extensive transmission rights-of-way. Loss of locational flexibility for electricity generating plants converting direct or indirect solar flows and inevitable land-use conflicts with food production would be additional drawbacks.

Such a change would be highly challenging even with a close match of conversion and utilization densities. Average power densities of roof-top solar conversions and typical household needs overlap, but heating the current stock of single houses solely with solar energy during the whole temperate latitude winter would be impossible without interseasonal energy storage – and without major modifications of residential neighbourhoods and energy needs.

Even ideally oriented roofs could not cover the peak daily needs and do not supply enough energy during cold but overcast days, and it would make little sense to install expensive energy storages in old and poorly-insulated houses. In a significant share of existing housing, rooftops are either suboptimally oriented, or are shaded by neighbouring structures and trees. A readily utilizable potential may be quite limited even in apparently suitable neighbourhoods.

Effective solar housing would require planned introduction of well-designed buildings, a process which could replace most of the existing housing stock only after decades of gradual diffusion. And, in spite of their large roofs, it would be even more impractical to solarize existing super-markets – and impossible to contemplate such a solution for high-rises in a densely built-up downtown. Even a very efficient supermarket located in sunny Arizona would need at least three times its roof's space for collector arrays and high-rise downtowns would have to be surrounded by a collecting ring up to at least ten times their size if flat plates should deliver all of the space-conditioning energy and photovoltaics all of the required electricity.

No major inhabited region of the Earth has mean annual insolation densities so low as to preclude significant utilization of these fluxes at least as sources of space heat and hot water, but the variability of seasonal and daily flows will limit practical conversions. Indeed, the overall efficiency of collecting, storing and delivering energy in active solar houses has so far been below the initial expectations. Consequently, instead of supplying the bulk of wintertime demand, many active solar systems have made only minor contributors (Shurcliff 1986).

Covering transportation needs by biomass-derived fuels would be even less practical. Even when choosing the most efficient option – cultivation of sugar-cane for distillation of ethanol done with a net energy gain of about 0.3 W/m^2 – production of more than 30 EJ (exajoules) of these fuels currently needed by the world's motor vehicles would require virtually all productive land in the tropics. Relying on corn in temperate regions would bring much lower returns. Even during the transition period, when the distillation would be subsidized by coal power, densities for ethanol as a stand-alone fuel would be no more than 0.2 W/m^2. Power densities of a fully solar operation (fuelling the machinery with ethanol, distilling with phytomass heat) would drop to about 0.04 W/m^2 and the American motor-vehicle fleet of 1990 would have needed roughly five times the total US farmland!

These examples show how the very low power densities common to all biomass energy conversions would not fit the final consumption needs even after major efficiency improvements: a US car fleet even three times more efficient than the 1990 mean could not be supported by crop-derived fuels alone. Calculations for wood-based methanol show similar results. Only a handful of land-rich tropical countries could fuel relatively small car fleets with ethanol from high-yielding sugar cane or cassava: to run global car and truck fleets on biofuels is impossible simply on the basis of exorbitant space needs.

Substitution of coke by charcoal in iron metallurgy shows a similar power density mismatch resulting in enormous space needs for tree cultivation. Charcoal is an excellent alternative for iron ore smelting from the chemical point of view – but its friability means that it cannot support heavy charges in modern tall blast furnaces. Fully solar society would thus have to first rebuild its iron-making infrastructure, and then it would face the limitations of low power density production inherent in wood production. Provision of enough charcoal just for the recent output of roughly 500 Mt of pig iron would call for some 3 billion cubic metres, or about 30 EJ, of wood.

Even if all of this wood would come from intensive short-rotation plantations producing annually about 5 t/ha (roughly 0.25 W/m^2), such planting would have to occupy at least 300 Mha, or an equivalent of nearly 20 per cent of the world's forests, an obviously impossible choice. The only plausible alternative to charcoal-based smelting in a solar civilization would be to use direct reduction processes powered by concentrated solar radia-tion: no such techniques were even in serious design stages in 1990. No smaller challenge would be posed by solely solar-driven synthesis of nitrogenous fertilizers.

These realities belie any simplistic notion that modern industrial civiliza-tion can rapidly shed its fossil-fuel foundations in favour of tapping, directly or indirectly, sustainable solar flows while not only maintaining the quality of life for the richest fifth of the mankind, but also bringing substantial improvements to the poor four-fifths aspiring for a better life. Of course, they also underscore the fact that doing without is, at least for the transitionary period of the next few generation, a more potent and a more desirable approach to transforming the world's energy base than relying on new primary conversion techniques. This brings me, finally, to that most amoebic category of limits to effective environmental management, to the enormities of socio-economic adjustment which will be required if we are to put the biospheric integrity above our ephemeral goals.

Enormities of socio-economic adjustments

In our minds we can map roads leading us toward a civilization preserving the biospheric integrity, but we simply cannot quantify the probabilities of arriving at the desired destination during the next few generations. What overwhelms us in the first place is the necessity to deal with so many separate yet allied challenges whose acuteness cannot be assessed with certainty, which cannot be ranked in any definite manner and whose management cannot be separated from a large number of fundamental socio-economic considerations.

Recent western focus on risks of climatic change, rising ozone levels, loss of biodiversity and tropical deforestation captures only a small fraction of the growing list of environmental challenges. To such chronic concerns as world-wide erosion and loss of farm soils, dumping of solid wastes, or pollution of streams, lakes and coastal waters – degradations which generally do not carry immediate catastrophic connotations – we must now add the clean-up of nuclear weapon facilities, an effort which will be among the most expensive environmental control undertakings.

Malfunctions, leaks and contamination from America's major weapon centres – Savannah River in South Carolina, Hanford in Washington and Rocky Flats in Colorado – brought attention to this problem during the 1980s. Preliminary estimates indicate that cleaning up and modernizing America's nuclear weapons complex will cost between US $(1989) 115 and 155 billion (Bowsher 1989). New information coming out of Russia indicates that the former Soviet weapons facilities have left behind an even costlier environmental contamination (Cochran and Norris 1991).

Similarly, recent prominent efforts to set up novel control mechanisms and to establish international co-operative frameworks are only a part of socio-economic realities – ranging from mass production of functional illiterates in schools to rising numbers of illegal immigrants – which will influence the outcome of our environmental efforts. These problems are always compartmentalized by bureaucracies in charge and, unfortunately, also by most of the researchers – but in reality they are interrelated in many inextricable ways.

And while the western perspective stresses conservation, clean-up and recovery, the poor world's focus for the next few generations will continue to be on the highest possible economic expansion, a trend inevitably resulting in more degradation and pollution. Resistance to lessons of proper conduct will be obvious: people who deliver them enjoy, on the average, incomes about a hundred times higher than those whom they see destroying unique environments and endangering the integrity of global commons. Effective world-wide environmental efforts will thus require not only an unprecedented degree of international co-operation, but they will be also unthinkable without an extensive transfer of modern techniques and management skills from rich to poor nations. Yet experiences of the past two generations are hardly reassuring.

Since the 1950s every rich nation has been maintaining large bureaucracies dispensing aid to scores of poor nations but the results have too often ranged from tragicomic (when several granting agencies compete with each other on a small Caribbean island) to disastrous (when aid has actually resulted in reduced labour opportunities and greater dependence on imports, or in accelerated environmental degradation). The most fundamental misunderstanding of the development industry has been the repeated ignorance of the total setting in which the aid is to perform, the inorganic introduction of modern parts into traditional wholes. And many of those settings are fundamentally inimical to consensual, co-operative progress. Those who think how efficient techniques and market incentives could transform the inefficiency and degradation in poor countries underestimate the depth of social malaise, the absence of rational discourse, and the unwillingness to compromise which prevail in many of those nations.

They would do well to think about the destruction of polity in Somalia or to read, for example, Roque and Garcia's (1991) depressing assessment of economic inequality and environmental degradation in the Philippines, Heng and Shapiro's (1986) tales of Chinese corruption, or Chileske's (1988) survey of Africa's socio-economic failures. What makes us think that these disgraceful arrangements will yield quietly to new consensual habits resulting in the introduction of critically needed economic and legal incentives which would prevent large-scale plunder of natural resources and enable a gradual restitution of local environmental integrity?

They should also check many readily available surveys of top arms importers – and find Vietnam, India, Egypt, Angola, Afghanistan and Ethiopia, all poor nations with badly ravaged environments – very high on the list (UNDP 1991). What gives us the reason to believe that these perverted priorities, these hatreds and enmities, will recede during the next few generations to make way for enlightened developmental and environmental management? And even the best-intentioned governments find the effectiveness of their actions limited by the paucity, often near-absence, of organizational and managerial infrastructures ranging from an adequate legal system and regulatory bodies to efficient tax collection. Building up these foundations is a protracted affair: as continuing manpower recruiting campaigns attest, even the richest of the Middle East have not succeeding in putting them in place after a generation of fabulous earnings.

But this criticism offers no absolution to the rich world. We have come up with clever debt-for-nature swaps, but Sheldon Annis was right when he concluded in his Senate testimony that 'it is the *nonswapped* 99.9% of the debt . . . that we really have to worry about' (Cody 1988). Repayment of these imprudently acquired debts will make it more difficult to narrow the world-wide gap between rich and poor nations. In 1960 the richest 20 per cent of global population claimed about 70 per cent of the world economic product; by 1989 that share rose to about 83 per cent. People in rich countries are now, on average, about 60 times better off than the poorest fifth of the mankind, the gap twice as large as it was 30 years ago! Can this be a basis for world-wide co-operation on assuring the biospheric integrity?

Even when we would try to forget this overhang of poverty, and look merely at the prospects of transforming the attitudes and actions in rich nations, we cannot find much encouragement. A common perception is one of a marginal fix, rather than of enormous socio-economic change. The rich world's bookstores, where environmental concerns used to be represented merely by large coffee table books about animals and by gardening manuals, feature a number of publications advising the public how to be green and how to save the planet. These publications deal with good intentions, and efforts on every scale will be essential in order to bring any effective change. Still, will the Earth really turn green when all of us will use such environmentally-friendly household cleaners as a silver polish prepared by boiling one litre of water with a tablespoon of baking soda and salt and a piece of aluminium foil (Getis 1991)?

Public opinion polls may show considerable concern about the state of the environment, but what would be the common attitudes once the people would be asked to pay more (indeed much more), to consume less and to discover the rewards of frugality, and once they would start realizing that such a state of affairs is not a temporary sacrifice but a permanent commitment to biospheric salvation? This adjustment would hit especially hard the world's most affluent nation. That the United States – the leading consumer of energy and raw materials and pre-eminent generator of pollution, as well as still the most dynamic modern society – must have a special role in fashioning a new environmental order is indisputable.

I have to cite an eloquent affirmation of this fact from a surprising source, Archbishop Desmond Tutu of Capetown:

> You are a powerful people. You can make the world a better place where business decisions and methods take account of right and wrong as well as profitability . . . You must now make a stand on important issues. Your decisions affect people – people who are of infinite worth because they are created in the image of God. God depends on you.
>
> (Tutu 1991).

This may be a captivating rhetoric – but is also a sobering reality. How much more worrisome is then the country's continuing economic drift, deepening social malaise and disappearing readiness of its ever more splintered special interests to compromise.

Perhaps the most basic obstacle to this is the frightening decline in learning: real understanding seems to be indirectly proportional to the enormous amount of information available in the American society. An overwhelming share of young people born after the mid-1960s grew up in households with at least one TV set, and they spent more time watching it than talking to their parents or reading. Actually, most of them have never really read, just some desultory compulsory chores – and when they grew up they were told to become computer-literate instead. What a ridiculous term: applied to young people who, measured by the standard of their educated grandparents, are almost illiterate. This crisis will not be readily remedied.

Prospects for readjusting government spending are not bright either. By 1990 all levels of American government were spending just over US $11,000 on every citizen over 65 years of age, compared to only US $4,200 on children under 18 (Smith 1991). America's seniors, 12 per cent of the country's population, were commanding just over 50 per cent of all federal social spending. This share, claiming almost 6 per cent of the country's GDP, is expected to approach 9 per cent during the next generation. The two largest expenditures – Social Security and Medicare – will be extremely hard to trim. Not only will there be rising real demand with larger numbers of old people, but the ageing of the affluent baby-boom generation will only increase the expected level of medical care. Where will this leave the education and health of America's children? Already single women with young children are the most rapidly growing group of homeless persons (Bassuk 1991), a trend which almost guarantees the growth of America's underclass.

At the same time, after decades of holding steady, America's middle class is shrinking as the work-force gets increasingly segregated into professional high-paying jobs and unskilled low-paying ones (Wallich and Corcoran 1992). And one must doubt that the powerful US business can carry its critical weight in this great transformation when even some of the largest American companies with enormous stakes in environmental management seem to be in chronic difficulties: General Motors' troubles come first to mind (Taylor 1992).

Generations of living with low costs of energy, water, food, wood and minerals created vested interests for perpetuation of these unrealistic prices among both the rich and poor consumers. How will the combination of poor education, massive functional illiteracy, skewed government spending, habituation to low resource costs, growing income gaps, family decline and homelessness, affect the prospects of adopting economically painful environmental strategies? What will it do to the chances of properly discounting future benefits? No other nation can fill the US leadership role. Japan has economic strength and technical know-how, but it lacks the appeal of universal political ideals, and its xenophobic habits and powerful bureau-cracy are hardly models for reconciling contentious international concerns (Schlosstein 1991). Moreover, the Japanese environmental record – except for the rapid and impressive clean-up of air pollution (Nishimura 1989) – is decidedly mediocre.

Prospects for substantial environmental advances are incomparably dimmer in the collapsed superpower. Remedying the environmental destruc-tion of the former USSR (Feshbach and Friendly 1992) can hardly be a priority for countries burdened by the legacy of seven decades of Communist mismanagement which demands straightening the economic basics – prices, markets, budgets, ownership, legal system – first in order to meet the existential needs, and then to start satisfying enormous pent-up expecta-tions.

Enormous adjustments involving major social and economic dislocations would also be needed in order to deal with the unprecedented challenge of equitable sharing of remedial burdens across the breadth of the devel-opmental spectrum: as all greenhouse gases are long-lived compounds and get mixed throughout the atmosphere, a truly global participation will be essential for the success of such efforts. Although the recent agreement limiting the global emissions of CFCs has been praised as a fine precedent (Wirth and Lashof 1990), I do not believe it is the case. That deal was facilitated by a combination of four facts: it benefited from a fear of an impending catastrophe, it did not have to set up complex regulations, it could rely on a technical fix as the principal solution, and it did not involve sensitive equity issues.

The undisputed conclusion that CFCs are linked with a massive (in excess of 50 per cent) seasonal destruction of stratospheric ozone above Antarctica, and fear of a similar rapid destruction of a large share of stratospheric ozone above the higher latitudes of the northern hemisphere, made the agreement a matter of obvious urgency. Instead of searching for a complex regulatory formula or looking for such unorthodox ways as meeting an agreed target by tradable national permits, the agreements simply call for the fastest possible elimination of the offending compounds. This goal is made possible by the rapid pace of developing better substitutes for CFCs (Baxter and Fairchild 1990; Manzer 1990).

None of these conditions will apply when trying to negotiate a treaty aimed at preventing or reducing the risks of global climatic change. Even if they would be relatively very rapid, the effects of global climatic change induced by emissions of greenhouse gases would unfold over many decades, and their inherent complexity would make it difficult to present the threat in unequivocally catastrophic terms. Consequently, even if everybody would accept the inevitability of notable global warming during the coming generations, all major environmental changes arising from this shift are disputable, if not in their general effects then certainly in their national impacts. And as there may also be appreciable benefits to some countries, nobody can offer any generally acceptable accounts of most likely net effects.

The second difference involves the extent of participation, and hence the complexity of possible agreements. The European acid rain treaty included only 30 nations, most of them with similar levels of income and technical ability, and while by 1992 there were about 60 contracting parties to the Montreal Protocol controlling CFCs, the bulk of these compounds is produced in only about two dozen nations (most of the world's more than 150 countries do not produce these compounds at all) – and all but two of these are highly or fairly affluent. This means that the agreement can work with relatively modest capital transfers, mainly for substituting Chinese and Indian usage.

In contrast, every country generates CO_2 from combustion and nearly all of them also produce CH_4 and N_2O from fossil fuels and agriculture – while technical fixes have only a limited usefulness in reducing their production, and while there is no realistic possibility of eliminating the causes of these emissions for generations to come. Rich industrial countries are the largest producers of greenhouse gases, but even their drastic unilateral actions would make little long-term difference. For example, if Germany, Europe's largest economy, would stop burning all fossil fuels, global CO_2 generation from coal and hydrocarbon combustion would drop by only about 4 per cent (Boden *et al.* 1990), a decline which may be filled by China's increasing use of coal in just a single decade.

And while CO_2 from fossil fuels still accounts for the largest share of greenhouse gases, many countries with extensive biomass (forest and grassland) burning produce more CO_2 per capita than most of the rich nations (Figure 5.6). An effective agreement reducing future emissions of greenhouse gases would have to be truly global, and it would have to address the contentious questions of equity, affluence, consumption rights and population growth. There are four principal strategies for sharing control costs: equal percentage reduction, ability to pay, polluter pays principle, and the natural right to emit (Toman and Burtraw 1991). An extension of the third approach would be to adopt cuts proportional to past cumulative emissions (Epstein and Gupta 1990). Everyone of these approaches would bring unacceptable economic and social burdens to some parties.

Equal percentage reduction of emissions (the strategy adopted by the 50 per cent club of 30 European nations to control acid deposition) could not equalize marginal control costs (not even if just rich countries would participate), and hence it would not be the least-cost choice. Its cost would be especially high for countries heavily dependent on coal. North America and Western Europe could certainly ride out a retreat from coal without nation-wide economic disruption, but, inevitably, there would be painful regional consequences in abandoned coal-mining areas.

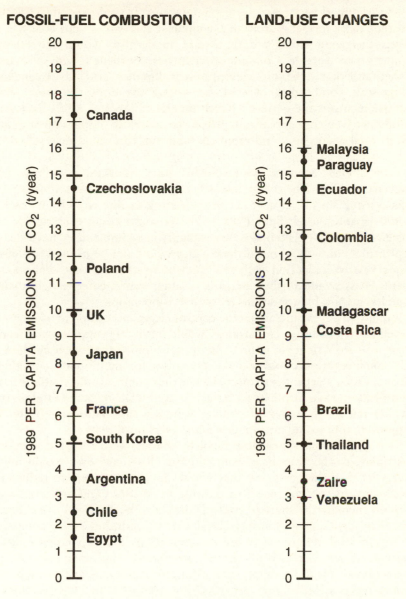

Figure 5.6 Although it is much more difficult to quantify CO_2 emissions from the burning of forest and grassland phytomass, estimates in the second column show that, in per capita terms, many poor countries now generate more of the gas through land-use changes than many rich countries release from fossil-fuel combustion.
Source: Plotted from data in World Resources Institute (1992).

But the world's largest per capita emitters of coal-generated CO_2 also include most former European Communist economies, and coal has also been energizing China's dash toward modernity. Would the relatively impecunious former Communist countries, burdened with a systemic inheritance that in the foreseeable future makes them critically dependent on large-scale combustion of coal, be willing to adopt the same relative emission cuts as the two rich North American nations? And China simply could not even contemplate any deep cuts: coal now supplies three-quarters of its total energy use, and rapid consumption cuts could reverse a decade of growth.

The Japanese economy, as in so many other instances, would present the obverse problem. The energy intensity of Japan's economy is already so low that even a 30–40 per cent cut in the American rate would not reach the 1990 Japanese level (Smil 1991c). The Japanese could thus rightly see a demand for equal percentage cuts as highly inequitable (they have cleaned-up their act already) – but if the country would not be pressed to share an equal relative reduction with less efficient western nations, then its competitiveness would further increase, creating even greater trade imbalance and less favourable conditions for global co-operation.

Ability to pay would assign the control targets according to average GDP levels in each country. This strategy would inevitably entail massive transfer payments from richer to poorer countries. I simply cannot envisage how any US administration could ask the taxpayers for multibillion subsidies to reduce China's dependence on coal while the country would continue having multibillion trade surpluses with the USA and while its human rights record would remain a matter of chilling Amnesty International reports. This approach, too, would not equalize marginal control costs.

The polluter-pays principle requires sharing of the agreed reductions according to the amount of emitted gases. Rich countries would initially carry most of the costs, but they could move aggressively to reduce their emissions. In contrast, needs of growing populations and industries would rapidly increase the share of control costs born by poor countries. Because the rich world was not burdened by similar expenditures during comparable stages of its development (when its populations and expanding industries accounted for virtually all global greenhouse gas emissions) such an arrangement could be rightly seen as discriminatory, as a strategy denying equal rights of economic progress. But trying to include penalties for historical emissions would be highly controversial: how far would such obligations go, would they be taken as aggregates or as per capita values? And this strategy, too, would not equalize marginal control costs.

In order to be equitable, allocation of rights to emit on a per capita basis would have to take into account historical differences between rich and poor countries – but such an adjustment would amount to a differential taxation and it could encourage movement of energy-intensive industries to poor countries. This tendency to relocate could be prevented by levying tariffs on their exports – but this could not be handled readily in a world where much simpler international trade agreements remain elusive. On the other hand, in order not to reward higher population increases, rich countries might want to put some limits on allowable claims in many poor countries with rapid growth. This would amount to externally imposed population control targets, a most unlikely possibility.

The natural right to generate greenhouse gas emissions could also be interpreted in such a way that China or India could not be blamed for any greenhouse gas emissions. This approach, advocated by the Indian Centre for Science and Environment (Agarwal and Narain 1991) argues that emission rights should be based on per capita shares of greenhouse gas sinks. If, say, 50 per cent of all emitted CO_2 is absorbed by oceans and plants, then the average global allowance in 1990 would be close to 600 kg C/capita. China's average in 1990 was just below that rate, India's only about a third of it – and so the world's third and sixth largest aggregate emitters of CO_2 are blameless. Needless to say, this kind of reasoning will not be accepted by rich nations. Even if accepted on moral or economic grounds, allocations would be made difficult owing to uncertain and changing capacity of known sinks.

Moreover, serious difficulties would start even before the selection of one of these contentious strategies, because we simply do not know the total amount of CO_2, CH_4 and N_2O emissions with satisfactory accuracy. Major uncertainties are due to emissions from biomass combustion, especially from tropical deforestation. Even if its total share would be only about one-tenth of total CO_2 flux it would be inequitable to expect coal-burning North China (where there is little left to deforest) to reduce its CO_2 emissions while ignoring Brazilian burning of the Amazonian rain forest. Potential unfairness becomes even more obvious when it is realized that average annual per capita CO_2 emissions in coal-burning Shanxi are now about 1 t of the gas, while a Brazilian peasant clearing just one hectare of Amazonian forest will be responsible for releasing about 500 t CO_2, an equivalent of ten years of per capita CO_2 emissions in the United States.

Molecules of atmospheric CO_2 are identical – but not from the perspective of poor countries:

> As a matter of fact, the North cannot escape from its direct and indirect responsibilities in the environmental situation of the South. The current situation of developing countries is, in general, a consequence of past colonialism, neocolonialism, and imperialism.
>
> (Dourojeanni 1989)

For rich countries this perception poses difficulties no matter if it is uncritically accepted, totally rejected, or taken as a basis for negotiation. The first choice makes the rich world responsible and dictates heavy financial responsibilities: this course will not enjoy widespread political support when domestic problems make people feel strongly about spending priorities. The second course puts an end to efforts aimed at effective world-wide co-operation. The third one runs into the already reviewed difficulties of valuation, poor understanding and unforeseeable outcomes.

At the beginning of this book I illustrated our difficulties in understanding the biospheric complexities by looking closely at fertilization of a field. Before turning to the last chapter, I would like to close this survey of limits to environmental management with another field example, with the story of sugar production. Its mishandling offers an outstanding example of linked complexities which make proper environmental management so elusive. Sugar's links – ranging from fundamental natural constants to irrational economic development, and from basic human equity concerns to world-wide ecosystemic degradation – demonstrate the limits of effective action with a depressing perfection.

Since the beginning of the sixteenth century, when the Caribbean plantations started to export sugar, until the early decades of the nineteenth century it was the single largest European import. But when Benjamin Delessert developed Marggraft's old idea of making sugar out of beets – an achievement which earned him in 1812 Napoleon's own impulsively given *Légion d'honneur* – it was only a matter of time before Europe's industrializing nations set up their large domestic sugar industries and started fixing sugar prices and restricting free trade. After nearly two centuries of such interventions today's sugar market is still riddled with ridiculous price fixing and trade distortions which are both very costly and which have enormous environmental impacts.

During the late 1980s the European Economic Community paid its farmers more than five times the world price of refined sugar – and the farmers responded with ever higher outputs which the community has been exporting at a large loss. Similarly, the US sugar producers were paid more than four times the world price, the resulting large domestic outputs cut the previously large imports and made alternative sweeteners, above all the high-fructose corn syrup (now the principal sweetener in soft drinks) artificially competitive. Without these interventions beet could rarely compete with sugar from cane. Cane's advantages come from fundamental differences in photosynthetic pathways encoded by the evolution.

Sugar-beets are C_3 plants, storing the intermediate products of photo-synthesis in a three-carbon acid, while sugar-cane is a C_4 plant, forming first four-carbon acids. C_4 plants have inherently higher productivity and more efficient water utilization than C_3 species. Sugar-cane's natural advantage is further increased by the fact that it is a tropical grass photosynthesizing all year, with the first two resprouted (ratooned) harvests yielding more than the first cut; then a decline sets in and the field is replanted (Blackburn 1984). Cane's global average yield is now close to 60 t per hectare per year, with the best producers averaging around 150 t (FAO 1991). With recoverable sugar content at about 10 per cent (7–15 per cent range), even an average hectare will yield some 6 t of sugar – and after the removal of the juice the remaining fibrous stalk residue (bagasse) can be burned to fuel the sugar-refining process which requires plenty of steam.

In contrast, the average global harvest of sugar-beets is about 30 t per hectare and with sugar content also at about 10 per cent the mean sugar yield is only half of that from sugar-cane. Production of sugar from beets cannot be self-fuelling and, owing to the crop's poorer photosynthetic efficiency, it requires relatively more fertilizer to sustain high yields. Simply, sugar in temperate regions can never be grown as efficiently as in the tropics. Moreover, the inexpensive labour force in tropical countries offers a clear cost advantage, refining of sugar in the producing countries opens the way to industrialization and its exports can provide foreign exchange earnings to generally indebted tropical nations. Instead, the rich world stimulates huge overproduction of its costly sugar which depresses the world price: during the 1980s it fell (in constant US $) to as low as 9 cents/kg, and by 1990 it was less than a third of the record level of the mid-1970s.

In order to benefit a very small number of farmers, rich countries have pursued policies which make little economic and environmental sense, and which affect relatively large numbers of the poor world's peasants. Removing, or at least greatly reducing, this kind of irrational arrangement would go much further in dealing with the linked problems of environmental degradation and economic development than staging international conferences. This example, contrasting the embedded irrationality of our practices and potential gains of better management, also foreshadows the closing chapter of this book which will focus on the *yin-yang* nature of environmental challenges: so difficult, yet with so many promising solutions, seemingly so intractable, yet amenable to effective changes.

6

POSSIBLE OUTCOMES

UNCERTAINTIES AND HOPES

Die Menschen sind wie Flugsand, und man ist nie sicher was morgen oben liegt.

Albert Einstein

Uncertain future is a key reality of human condition. We strive for a greater control, and during the past generation we have invested a great deal of intellectual effort into improving the capability of our quantitative techniques, and produced increasingly complex long-range forecasts. But retrospectives show little real success. As always, unpredictable discontinuities provided the best antidote against over-ambitious forecasting efforts. They had profound effects in every sphere of human action: from the Yom Kippur War to the disintegration of the Soviet empire, from floating exchange rates to widely fluctuating commodity prices, from the fall of western labour unions to the rise of Muslim fundamentalism, from the global primacy of Japanese car-making to the microcomputer explosion, and from the discovery of chlorofluorocarbon (CFC) effects on the stratospheric ozone to fears of global warming.

There is no reason to assume that frequency of such profound unpredictable changes will diminish in the future. And even should we be successful in forecasting fairly accurately many particular developments, it is most unlikely that we will be able to foresee the complete settings of such changes. Such wholes matter much more than their parts, and forecasting the wholes of civilizational development is far beyond our abilities. An outstanding example suffices to illustrate this common impossibility.

In 1970, when China was slowly emerging from the madness of the Cultural Revolution, and when it had not published any proper statistics for more than a decade, a western demographer could have been lucky by forecasting quite accurately the country's 1990 population total of nearly 1.15 billion people – and this total could lead to speculations about China's desperate food and economic prospects. But how many experts in 1970 would have predicted that by 1990 Chinese farming would be entering the second decade of privatized production, and that it would be producing enough food to bring the country's average per capita supply within less than 10 per cent of the Japanese mean? Or that China's economy would be the fastest growing one in the Asia of the 1980s, generating a large trade surplus with the United States?

And so I will not offer yet another set of elaborate but almost instantly irrelevant forecasts. The most robust look must concentrate on those inevitable realities and trends whose outcomes cannot be fundamentally different regardless of the intervening discontinuities and altered wholes. I will limit my look to no more than two generations: a combination of demographic imperatives and operational inertia of principal energy and material supply systems guarantees that the shape of the world up to about half a century ahead is far more discernible than the more distant developments. An even more important particular justification is that the next two generations may well be the most critical period in the history of industrial civilization, and they may rank among the watersheds of human evolution. Why this may be so is best appreciated by viewing these decades in an evolutionary setting.

TWO PERSPECTIVES

On the other hand, the internal perils that life has to face, arising from the emergence within it of a reflective liberty ... seem much more menacing and alive.

Teilhard de Chardin, *Man's Place in Nature* (1966)

Life on the Earth has been evolving for more than three billion years and astrophysical imperatives allow for almost twice that time for further complexification. The inevitable terminal catastrophe awaits about five billion years from now, at a time when our star will exhaust a large part of hydrogen near its centre and the domain of the fusion reaction, converting hydrogen to helium and keeping stars alive, will travel rapidly outward, heating the Sun and expanding it enormously (Friedman 1986). Photosphere, the Sun's outermost layer now about 150 million km distant from the Earth, will extend beyond the orbit of Mars as the giant star will vaporize all the terrestrial planets.

If no other catastrophic events – extraterrestrial or geogenic – would destroy it, a truly sapient global civilization could be thus evolving on the Earth over a span a million times longer than the time elapsed from the building of the first pyramid or the rule of the mythical Yellow Emperor (both events date to about 2800 BC). Other extraterrestrial catastrophes affecting the Earth are, of course, possible during the coming billions of years – but none of them is inevitable and their probabilities are exceedingly small. Perhaps the least likely is a 'nearby' explosion of a supernova which would flood the Earth with doses of radiation high enough to extinguish all life (Cosmovici 1973).

In spite of up to ten postulated episodes of exposures to 500 roentgens (a dose lethal for most mammals) since the Precambrian period (600 million years ago), actual radiation doses were too low to derail the course of evolution and the future probabilities of destructive impact on civilization can be further lowered owing to the forewarning to be expected before the arrival of radiation. If it is assumed that an advanced civilization could prepare shelters for its population during the year elapsing between the arrival of light and cosmic radiation, then an unavoidable dose of 500 roentgen should be encountered only once in one billion years. Probability of radiation catastrophe destroying much (most?) of human civilization would be then almost comparable to the unavoidable demise of the planet amidst the expanding red Sun.

Collision between the Earth and a large celestial object is an entirely different matter. Such currently uncontrollable encounters have undoubtedly happened in the past. A typical comet has a mass equal to a sphere of solid material with a radius of 10 kilometres (km) and in a head-on collision with the Earth the two objects could move with a relative speed as high as 70 km/second (Smil 1991a). But chances of a head-on collision are not higher than about once in a billion passages for any comet coming near the Earth, or about once in 100 million years. A substantial piece of cometary material may hit the planet once every ten million years. Past encounters may (or may not) have caused large-scale species extinctions (Alvarez and Asaro 1990) but they could not sever the evolutionary process. Nor could grand tectonic motions separating and rejoining the continents, periods of intense volcanic activity, or pronounced climatic changes (Kasting 1989; Sharpton and Ward 1991).

Until the middle of the twentieth century no human action could pose even a remotely comparable threat to the integrity of the biosphere. Then the possibility of full-scale nuclear war became our dominant existential concern. Mutual nuclear overkill acted as an effective restraint, and the disintegration of the Soviet empire has dramatically lowered the risks of nuclear war. But extremely low probabilities of extraterrestrial catastrophes and lowered chances of nuclear war do not add up to a nearly risk-free future for the biosphere. Two unprecedented developments – rapid growth of the global population and its quest for material affluence and higher quality of life – have combined to put such stresses on the biosphere that we are now increasingly concerned about its ability to support long-term development of modern civilization.

These concerns may be exaggerated, as the biosphere may prove to be much more resilient, even self-equilibrating (Lovelock 1979) – but they may also be quite inadequate, as different environments may be responding to human stresses by increasingly non-linear deterioration. The same uncertainty marks our ability to deal with such degradations. According to the catastrophists they will bring us much closer to an (irreversible?) civilizational collapse, while techno-optimists see in them opportunities for greater innovation. I must agree with B.J.L. Berry (1988) that 'at this juncture the choice is not one of science but of ideology'.

What is undeniable is the enormous extent of human effects on the biosphere, the reach extending from myriads of local changes to potentially destabilizing interferences in grand biogeochemical cycles. These realities demand responses – and the next two generations may be critical in formulating them and putting them into effect. This need seems to be even more urgent when we turn to the second perspective, to the consideration of critical demographic and socio-economic trends which will almost certainly mark the coming decades. First is the inexorability of further large population increases. Even if fertilities would be declining somewhat faster than expected, absolute population growth during the years 1990–2030 will surpass the record increase between 1950–90, and the global total will approach nine billion.

Second, more than nine-tenths of this increase will come in poor countries whose population expect to do better, but merely doubling the poor world's average standard of living over a period of four decades (leaving it still far behind the affluent mean) would call for much higher flows of fuels, food and raw materials. Simple arithmetic shows that even if intervening technical advances would increase typical energy conversion and material utilization efficiencies by 50 per cent, the combination of the poor world's population total about 70 per cent higher in the year 2030 than in the year 1990 and doubled per capita consumption would call for nearly doubling of energy and material inputs This would not be accompanied by inevitable intensification of biospheric degradation only if the rich world would be willing to give up a requisite share of its claims on biospheric resources and services.

Third, it will be impossible to accommodate the growing populations in rural areas. Inevitable intensification of farming will lower direct labour needs, and shortages of arable land will make its further subdivision increasingly difficult. High population and industrial densities accompanying massive urbanization will generate wastes at rates which even rich nations find very difficult to control. And high power density demands of large cities will make it much less likely that renewable flows will ease the environmental impacts of fossil fuel combustion. Fourth, while advances in crop breeding and agronomic management should make it possible to feed the expected population, those efforts will require further conversion of forests and grasslands to crop fields and further intensification of farming. Given the already existing soil erosion and declines in soil organic matter we must be concerned about the integrity of future agro-ecosystems.

And even if today's share of total photosynthetic production appropriated by humans is considerably lower than some published estimates, it is certainly already high enough to preclude another doubling without major global effects on water, carbon and nutrient cycles. Similarly, uncertainties surrounding capacities of biospheric sinks and atmospheric responses make it impossible to offer any accurate forecasts of global warming, but physical imperatives are certainly increasing the risk of relatively rapid warming: again, it is most unlikely that we could double the current rate of emissions of greenhouse gases without any significant biospheric consequences.

During millennia of civilizational growth we have destroyed and impoverished numerous ecosystems on local and regional scales. During the next two generations we may come close to compromising the biosphere's capacities to support further economic growth. If this is indeed the case – and if we act as risk-minimizers there is enough evidence to clearly assume so – then the next two generations will have to be a period when we will pay an unprecedented attention to maintaining the biospheric integrity, a task which will necessitate a transition to fundamentally different economic arrangements. I believe that the resulting socio-economic transformation would equal in its far-reaching implications the as-yet-unfinished worldwide transition from subsistence farming to urbanized industrial civilization. I also believe that we should approach this challenge with great humility.

PLANETARY MANAGEMENT

> As for those who would take the whole world
> To tinker it as they see fit,
> I observe that they never succeed . . .
> For indeed there are things
> That must move ahead,
> While others must lag . . .
> So the Wise Man discards
> Extreme inclinations
> to make sweeping judgements . . .
>
> Lao Zi, *Dao De Qing*

These lines – traditionally ascribed to Lao Zi, a recluse who became a historian in the secret archives of the Zhou dynasty – are more than 2000 years old and they contain the wisdom of a civilization which at that time had already accumulated several millennia of tempering experience with managing floods, droughts, invaders, interstate feuds, public duties and private probity. And Lao Zi's sentiments would have been comprehensible to more than one hundred successive generations that lived with the all-embracing obligations and promises of traditional order, in China or elsewhere.

Only the near-magical advances of modern civilization led to the rapidly widening circles of tinkering, and to often arrogant perceptions about the chances of its success. Now, in order to preserve the biosphere capable to sustain further socio-economic advances, we face the challenge on the ultimate scale: managing the planet Earth. Such a task may appear to be a clear necessity, but the mission should make us very cautious. We should be highly critical of any sweeping judgements dominating discussions of environmental change, and we should be humbled by a large number of lesser challenges where we have tinkered – and failed as nations or as the international community. Yet even people who should know better are repeatedly drawn to sweeping judgements.

Are ecosystems fragile or resilient? Fragile seems to be the only adjective encountered in standard writings on environmental change. But before using the term one must define the criteria of fragility and examine their temporal trends in particular ecosystems. And while it certainly describes many properties of many ecosystems, fragility is not a general attribute of all ecosystemic functions in all communities. Will the anticipated global warming be fast or will the rate of change be tolerable? 'Very fast' reply the proponents of the catastrophic change who believe that the warming will bring 'starvation, poverty, squalor, and streams of environmental refugees at a level which we can hardly imagine today' (Topfer 1991). But certainly slower than the change in critical components of economies adjusting to slightly higher temperatures. Besides, 'fast change that is anticipated may matter less than slower change that is unforeseen' (Ausubel 1991b).

What kind of management will successfully bridge these sweeping judgements? And how does one know that the population of a particular country has already surpassed the carrying capacity of its environment? Carrying capacity is not too difficult to define for deer or gorillas – but without detailing average energy and material flows it is an enormously elastic concept for human societies, and one made even more fuzzy by increasing international trade. Only with clear definitions can we get meaningful answers, but there is little chance that undefined, but highly evocative, terms will disappear from the discourse of environmental change.

Similarly, it is not easy to judge the impact of environmental degradation on people. I find it impossible to believe that greater crowding will make for a higher quality of life (Simon and Kahn 1984), but it is quite obvious that a great deal of environmental degradation is acceptable by people as long as their material standard of living is rising. Indeed, as Day (1971) noted, the fluid nature of human values is a major threat to any effort blocking gradual deterioration of environmental quality. During the 1980s undeniable improvements in the average standard of living in many poor countries were embedded in the generally declining quality of their environment. Of course, there is nothing new in this trade-off: Europe and North America lived with it for many generations before the 1960s.

The only rational conclusion is that the complexity of links between population growth, resource use, environmental quality, and innovation leaves little room for simple generalizations. Closer looks almost always reveal a peculiar combination of disheartening trends and encouraging changes. Contrary to the commonly held impression (another case of an inexcusably sweeping judgement), Thomas Robert Malthus, the great initiator of modern debates on population growth and resources, was a thinker espousing such an eminently realistic position.

Of course, he will always be best known for his conclusion that 'the power of population is indefinitely greater than the power in the earth to produce subsistence for man' and that 'this natural inequality ... appears insurmountable in the way to the perfectibility of society' (Malthus 1798). But his was a more complex vision, eloquently set in the second edition of his great essay (Malthus 1803):

> On the whole, therefore, though our future prospects respecting the mitigation of the evils arising from the principle of population may not be so bright as we could wish, yet they are far from being entirely disheartening, and by no means preclude that gradual and progressive improvement in human society ... And although we cannot expect that the virtue and happiness of mankind will keep pace with the brilliant career of physical discovery; yet, if we are not wanting to ourselves, we may confidently indulge the hope that, to no unimportant extent, they will be influenced by its progress and will partake in its success.

But a confidence in such an outcome is weakened by the burden of our failures. Too many countries in the poor world have been crippled by racial hatreds and killings, civil wars, mass corruption and organized crime, starvation, malnutrition, and extreme poverty – and in some of them these apparently intractable factors dominate the outlook. While this social disintegration does not invalidate concerns about environmental integrity, it certainly creates a different order of priorities and it reduces, even eliminates, meaningful participation in any global effort.

Dissipating forces are also at work in rich nations, where the challenges of racial discord, deep income disparities, economic stagnation, budget deficits, crime and drug addiction, mass functional illiteracy, poor education, family breakdown and public apathy combine to create conditions hardly conducive to pushing resolute policies elevating environmental management to a national priority. How readily will even these countries agree to long-term participation in global agreements specifying their costly commitments to the maintenance of the global commons?

The grandest of these co-operative approaches would be to agree on a global compact, a bargain between rich and poor countries. Its essential trade-off would be a drastic reduction of energy and material use in the rich countries in return for the commitment to an early achievement of replacement fertilities in the poor world. For example, in Holdren's (1990)

208

scenario rich countries will halve their per capita energy use by the year 2025 (largely by deploying highly efficient techniques) so the poor ones could double it during the same period with the total world energy use increasing to no more than twice the 1990 rate assumed to be affordably and sustainably manageable with advanced supply techniques. This exchange would be accompanied by a programme stabilizing the global population at nine billion, a goal requiring the world-wide achievement of replacement fertility by the year 2010.

Imposing such clear targets may seem to be an essential goal of planetary management, but uncertainties of our understanding argue for their avoidance. Risks of global warming provide an excellent illustration. During the 1970s forecasts assumed long-term primary energy growth at over 4 per cent a year (the rate prevailing since 1945), and close to 60 per cent of CO_2 retained in the atmosphere. This would have doubled the pre-industrial CO_2 mean before the year 2030. But a substantial drop of primary energy growth rate during the 1980s, and its long-term continuation (roughly 2 per cent a year), coupled with an atmospheric CO_2 retention rate of around 40 per cent, would not bring the doubling of pre-industrial CO_2 levels before the year 2075. Demise of the USSR, the world's second largest consumer of fossil fuels, is bringing further declines of energy consumption and this could push the doubling date closer to the year 2090.

From an adaptive point of view the difference between the two outcomes is fundamental. A doubling taking nearly a century would offer many more opportunities for adjustments and innovations than one accomplished in half of that time. Of course, analogical arguments are valid for the preservation of biodiversity or for drastic measures to reduce soil erosion: we simply do not know enough to start imposing clear targets. But successful actions reducing or preventing environmental degradation do not have to wait for the adoption of firm goals, or for a formulation and acceptance of grand strategies arrived at by consensual global adherence to sustainability goals.

Using again the example of global warming, effective long-range national policies should aim at a significant reduction of greenhouse gas emissions regardless of the imminence or the intensity of global climatic change and the success or failure of international negotiations. This risk-minimizing strategy would be an insurance against the prevailing uncertainties and the possibility of dramatic surprises – but this benefit would not be the only, and not even the most important, reason for its adoption. Lower emissions of CO_2 would require reduced consumption of fossil fuels, a goal which we should have been pursuing aggressively all along because of its multiple benefits.

These gains include not only a large number of environmental improvements – ranging from the avoidance of land destruction by surface coal-mining to lower emissions of acid-forming gases, from reduced chances for major oil spills to cleaner air in urban areas, or from improved visibility to reduced disposal of fly ash and desulphurization sludge – but also notable socio-economic benefits. By reducing total exposure to a variety of pollutants lower fossil fuel combustion should bring decline in morbidity and longer life expectancy: these changes would not only improve the collective quality of life, but they would also affect health care costs. Lowered energy intensity of economic output would increase a nation's competitiveness in foreign markets, a development benefiting the balance of payments and creating new employment opportunities.

If a nation is a major importer of fossil fuels (and with the exception of Russia and China all large economies are), then reducing the dependence on those imports will not only improve its fiscal position, but it will also enhance its national security. And to achieve substantial reductions in fossil fuel combustion a nation will have to engage in a long-term development and diffusion of new techniques and management approaches, a development increasing its scientific and technical prowess and its managerial abilities and requiring better education: socio-economic spin-offs of these achievements could be as important as the efficiency gains themselves.

Clearly, our decisions to optimize the use of fossil fuels need not be based on the absence or presence of climatic threats as every change aimed at reducing fossil fuel combustion has a large number of independent, and ultimately perhaps even more important, benefits. Similarly, we should not extol the value of unknown genetic treasures in order to preserve biodiversity, and we do not have to wait for new national accounts internalizing the costs of species loss. In the longer run bioengineering advances may make any benefits of vine saps or insect-produced fungicides look rather modest, and even the best environmental accounting may still include many arbitrary assumptions. But a species-rich forest should have been valued all along for its essential environmental services which may be ultimately more important for the civilization's long-term survival than the demise of some endemic species.

Countries reducing their fossil fuel combustion or protecting their forests will thus have no regrets if there will be no pronounced global warming during the twenty-first century, or if advanced bioengineering will devalue their genetic stocks. No-regret strategies are clearly by far the best approach to environmental management. They move the countries in particular directions – controls, conservation, protection, efficiency, innovation – only if such commitments can be justified by more than an uncertain assessment of possible worst-case environmental risks and costs.

This pragmatic choice takes out most of the contention from the formulation of effective policies, and it avoids the difficulty of sustaining public and political support for measures with uncertain and remote benefits. But given the interconnectedness of biospheric changes, this approach does not leave out any major environmental concerns, and it should make it easier to select our management priorities by favouring measures with the largest number of identifiable environmental, social, economic and political benefits.

A great advantage of no-regret strategy is that it needs no grand agreements in order to make the difference. Energy and water conservation, higher conversion efficiencies, farming without excess, recycling, or preservation of ecosystemic services should be pursued as a matter of clear self-interest by all responsible nations without waiting for enactment of global accords which may be then seen anyway as too vague or too intrusive. There are two other notable advantages to this approach: it is necessarily incremental but many of its benefits can start flowing immediately; and as it is decentralized, it would avoid design and implementation perils of grand schemes and massive bureaucracies. Widespread, diffuse incremental adaptations are preferable to relatively sudden, centrally-ordered interventions whose frequent irrationality born of the crisis attitude is demonstrated by their rather rapid self-destruction, or by their notorious inefficiency.

Do you remember our grand government-sponsored solutions to energy crises? Our commitment to fast breeder reactors, corn-derived ethanol or to gargantuan synthetic fuel schemes? And how effective have been the huge and enormously expensive social-assistance programmes in reducing poverty and creating lasting employment? Do we want to commit analogical environmental blunders on a global scale by using uncertain claims in order to push dubious policies – or do we prefer to do things right because we will benefit in a number of important ways?

The fact that the industrialized countries in general, and the USA in particular, have been reluctant to do things right – that is, above all, to pay the real cost of extracting and converting natural resources and benefiting from environmental services – does not promise any rapid future changes. But this inexcusable record of irrational behaviour cannot simply go on. We cannot keep on encouraging high consumption and inefficient use of energy, water, food, feed or wood with low commodity prices – and then to suffer the impacts and cover the high costs of extensive environmental, health, social and military externalities created by such wasteful uses.

What will set the tone of world-wide environmental management will be the actions of rich countries, rather than communiqués of international conferences. Acting in their own self-interest, the rich countries must finally start pricing natural goods and services right. This will not be an easy transition to undertake, but there is hardly any more efficient way to finally recognizing that our economies are merely elaborate subsystems of the biosphere, and that the only way to ensure their long-term viability is to maintain environmental integrity. Inevitably, the burden of global leadership is on the rich countries. Can we look with some confidence to a relatively early start of the transition to saner economic arrangements?

CONFIDENCE OR DESPAIR?

> In the morning sow thy seed,
> and in the evening withhold not thy hand;
> for thou knowest not which shall prosper,
> whether this or that, Or whether they both
> shall be alike good.
>
> *Ecclesiastes* XI:6

Aggregate evidence of our limited understanding, unlearnt lessons, unrealistic expectations, slow adjustments and management failures is a rich source of exasperation and depressive feelings, especially when compared with the enormity of future challenges. But there is a profound distinction between these feelings and despair. Depression does not necessarily imply hopelessness about the future, while despair involves hopelessness as well as absence of larger human connection. 'Moments of depression are inevitable; they are part of being human . . . Despair is overcome when we begin once more to believe in, and act on, a human future' (Lifton 1985).

Effective actions must be based on our continuing progress in understanding the complexities of environmental change, but deeper knowledge alone will not suffice. Environmental realities are inseparable not only from existential necessities, but also from moral imperatives. Indisputable utilitarian success of the symbiosis between modern science and free enterprise is fundamentally self-destructive if not governed by the humility of a grander perspective. This need can be expressed in many different ways. One of its most eloquent confirmations comes from *Centessimus Annus*, the encyclical letter issued by John Paul II in 1991:

> Man thinks that he can make arbitrary use of the earth, subjecting it without restraint to his will, as though it did not have its own requisites and a prior God-given purpose, which man can indeed develop but must not betray. Instead of carrying out his role as a co-operator with God in the work of creation, man sets himself up in place of God and thus ends up provoking a rebellion on the part of nature, which is more tyrannized than governed by him.
>
> (John Paul II 1991)

The rich western world has obviously a unique responsibility in redressing this imbalance. Western civilization has had a global influence greatly disproportional to the number of people inhabiting its historic core and its new overseas outposts. The level of understanding reached by western civilization may take it beyond others in its fundamental behaviour, and it can become the foundation of a world civilization which, unlike any of its predecessors, 'might be perpetually reconstituted and maintained at a level undreamed of in past history' (Melko 1971).

Undoubtedly the single most important attribute which must be preserved in order to succeed in this challenge is our heritage of freedom, of dissent and choice. This is a liberating burden, a path to enormous opportunities beset with perils of wrong choices and grievous failures. There can be no guarantee of success, but profound scepticism, sustained inquisitiveness and uncircumscribed criticism are the best possible protectors of flexibility without which there is no viable adaptation and long-term survival. Scientific advice, now a critical part of this adaptive approach, must be as much a target of this questioning as our dysfunctional social and economic arrangements.

Will we be able to summon this intellectual vitality again and again – instead of just tiring out and fading away as other seemingly so successful civilizations? There is nothing inevitable about such a demise. The idea of societies as organisms going through predictable life stages from formation to decay is overly deterministic. There is no uncertainty in it but also no hope, no possibility of choice. Civilizations are not simple organisms with pre-ordained life cycles, but rather complex interactive systems able to adapt. Influences favouring their demise – environmental changes, external threats or internal malaise – can be buffered by social and technical adjustments and innovations (Butzer 1980).

Neither 'senility' nor 'decadence' are the causes of civilizational break-down: these most complex of all human systems meet their demise owing to concatenation of mutually reinforcing negative influences which overload the system beyond its adaptive capability. Civilization's survival remains largely a matter of maladaptive probability rather than of a cyclical inevitability. The challenge is thus concentrated in meeting two classes of complex demands: minimizing the probability of negative, mutually-reinforcing concatenations, and maximizing the capability for adaptive responses.

What needs to be done? Given the large number of readily effective particular measures and realistic prospects for further substantial technical and managerial advances, we must assure above all that basic conditions are right for adoption and diffusion of such adjustments. Nothing is then more important than opting for flexible strategies growing from vigorous free dialogue. Glimpses of free societies may be disquieting. Their continuous discords, their accentuation of material possessions and bad news, their often frivolous preoccupations and apprehensions about the future add up to an image of a tacky instability. To be sure, all of this needs much constant improvement, but this uncertain, unruly, seemingly directionless assemblage of contending interests wary of the future is also still the best known self-correcting adaptive arrangement of human society – precisely because of its low level of dogmatic commitment and relatively high, creative flexibility.

Anything diminishing these two great fundamentals of a free society must be then minimized to the greatest possible extent. This means above all the avoidance of any demands for the safety and security of fixed truths. Their public appeal may become commensurately higher with the difficulties and costs of many choices we will have to take during the next few generations in order to prevent intolerable levels of environmental and socio-economic decline.

We have to resist the adoption of sweeping normative solutions no matter from which part of the ideological spectrum they come, no matter which preconceived salvations they represent. The idea of creating a sustainable world through a conversion to a small-is-beautiful paradise should not be treated with less scepticism than a call for a sustainable civilization based on mass deployment of nuclear fission. Extractive uses of tropical rain forests should not be seen as inherently superior to well-managed logging contributing to a nation's real economic advance. We have to refuse, again and again, any offers of single-vision salvation: these dogmatic commitments would make it much more difficult to accommodate unforeseeable changes, their seeming simplicity would be negated by counterintuitive complications, their rigidity would foreclose precious flexibilities.

These would be the ways of simplifying maximalists, not the solutions of complexifying minimalists which are the best guarantee of flexibility and adaptability. To avoid inflexibility and simplification we have to engage in vigorous exchange of ideas. This requires constant intellectual vigilance as even science is prone to paradigmatic rigidities which may be hindering promising solutions. Indeed, the central question for the fate of mankind may be the rebirth of a genuine dialogue (Buber 1937). Such a dialogue can flourish only in free societies as they continuously, although often laboriously and discordantly, translate the intellectual flexibility into practical action and create futures which even the best minds are unable to visualize.

Management of environmental change is fundamentally the matter of human excesses, that is, the management of demands. Consequently, there can be no finer strategy than to avoid these demands or at least to minimize them, rather than to let them grow through neglect or inattention, or even to encourage them through irrational pricing and fiscal policies, and then to step in with regulations, taxation, controls, technical fixes and international treaties. Once this basic premise is accepted, two concerns become paramount: expanding populations in the poor world and the rising affluence of the rich nations.

Continuing rapid population growth in poor countries creates extremely burdensome demands. When Europe embarked on modern industrialization (1750–1800) its population growth averaged 0.6 per cent a year; a century later (1850–1900), during the decades of the greatest industrial expansion, the continent's population was growing by about 0.7 per cent a year (Demeny 1990). In contrast, between 1950 and 1990 Africa's population grew by 2.6 per cent a year, Latin America's just slightly less, and Asia's (outside China and Japan) by 2.3 per cent.

How could it be argued that such rates of population growth have little to do with pulling the people out of poverty, and hence out of their environmental morass? To wait for the expected feedbacks to work, that is, for the economic development to eventually lower fertilities to replacement level, is to condemn large numbers of living, and even larger numbers of yet to be born, Africans, Latin Americans and Asians to a degrading existence. Advances in food production may keep them just above starvation, but not without severe environmental cost brought by further intensification of cropping or by greater expansion of cultivated land. Inevitably higher urbanization will generate even higher energy demands and more concentrated pollution.

In terms of dignified material and intellectual gains this developmental path is for most of the affected people a reprehensible cul-de-sac. We know what works to reduce high fertilities – maternal education above all, ready availability of contraceptives, provision of basic infant health care – and we know that the benefit/cost ratios of these measures make them the best developmental, and that is also environmental, investments we could make. Clearly, vigorous population control measures are in the best interest of the worst affected countries.

Rich nations, with their near-stationary populations, should be no less aggressive in managing their energy and material demands by attacking both the frivolous consumption and enormous conversion and utilization inefficiencies. Realistic pricing is undoubtedly the best way to limit demand and to use resources efficiently. We should try hard to quantify the externalities of energy, food, water and raw material consumption and to pay for the real costs of these commodities. But this preference for market solutions is not incompatible with the advocacy of government interventions. Effective environmental management cannot do without realistic regulation, taxation and international agreements, and doctrinaire opposition to such approaches will ultimately be counterproductive. These adjustments, market or interventionist, may eventually yield substantial net economic benefits, but it is certain that they will cause a great deal of both temporary and permanent socio-economic dislocations. To portray such a great transition as a cost-free opportunity, or at worst as a penny ride, is irresponsible.

Difficulties abound, but possibilities exist everywhere. In rich countries they range from changing typical diets (a shift which would bring both notable environmental and health benefits) to getting serious about mandatory recycling of most usable waste materials. Of course, by far the greatest environmental gains can come from reduced energy use, and detailed sectoral studies show effective solutions even for the extraordinarily high and structurally embedded American energy addictions (OTA 1991).

Effective solutions could be found even in seemingly hopeless cases. Meier and Quium (1991) describe a path which, without any arcane techniques, could make the living in Dhaka, the Bangladeshi capital, and one of the world's poorest cities, considerably better by the middle of the next century.

215

Fitjer (1990) notes that a major share of Ethiopia's impressive hydro energy capacities can be developed by small rural stations: as in China, they would act as affordable, decentralized energizers of rural modernization. And a combination of traditional agronomic practices and bioengineering advances could turn the trend of African food production.

Challenge of the transition, enormous as it is in its scope and unique as it may be in its rapidity, is still certainly within our abilities. Any open-minded survey of civilizational transformation accomplished during the twentieth century must end up not only with exasperation, regrets and despair – but also with elation, approval and hope. Nobody can predict what our chances are in the very long run – but in spite of the uncertainties in our understanding of environmental change it is most unlikely that human actions have driven the biosphere on to a slope of an irreversible catastrophic decline. Urgency of decisive remedial actions will increase appreciably during the next generation, but today the outcome is as open as ever. The title of this closing section has thus a doubly negative answer. Neither confidence, nor despair.

In his profound collection of essays on the tragic sense of life Miguel de Unamuno (1921) formulated the answer with a persuasive recourse to one of his country's greatest literary heroes. What is needed is 'a faith based upon incertitude . . . of such kind was the heroic faith that Sancho Panza had in his master, the Knight Don Quijote de la Mancha; a faith based upon incertitude, upon doubt'.

One cannot judge civilization's record without deep exasperation, and the magnitude of challenges does not allow any simple, reflexive optimism. But it should not lead to despair either. Writing about the fears of nuclear conflict Robert Jay Lifton (1985) noted that:

> We need to be neither optimistic nor pessimistic about the human future; rather, we must hope. Hope means a sense of possibility and includes desire, anticipation, and vitality. It is a psychological necessity and theological virtue, a state that must itself be nurtured, shared, mutually enhanced.

Earlier in this century Miguel de Unamuno (1921) expressed the same need even more forcefully: '*Spero quia absurdum*, it ought to have been said, rather than *credo*'.

How can one disagree?

REFERENCES

Abelson, P.H. (1991) 'Excessive fear of PCBs', *Science* 253: 361.

Adams, W.M. (1990) *Green Development: Environment and Sustainability in the Third World*, London: Routledge.

Ad hoc Committee on the Relationship between Land Ice and Sea Level (1985) *Glaciers, Ice Sheets, and Sea Level: Effect of a CO₂-induced Climatic Change*, Washington, DC: US Department of Energy.

Agarwal, A. and Narain, S. (1991) *Global Warming in an Unequal World*, New Delhi: Centre for Science and Environment.

Ahearne, J.F. (1989) 'Will nuclear power recover in a greenhouse?' *Resources* Winter 1989: 9–11.

Aldrich, S.R. (1980) *Nitrogen*, Urbana-Champaign, IL: University of Illinois Press.

Alvarez, W. and Asaro, F. (1990) 'An extraterrestrial impact', *Scientific American* 263 (4): 78–84.

Anderson, A.B. *et al.* (1991) *The Subsidy from Nature: Palm Forests, Peasantry, and Development on an Amazon Frontier*, New York, NY: Columbia University Press.

Anderson, R.M. and May, R.M. (1992) 'Understanding the AIDS pandemic', *Scientific American* 266 (5): 58–66.

Andrews, F.M., (ed.) (1986) *Research on the Quality of Life*, Ann Arbor, MI: University of Michigan.

Arnold, F. and Liu, Z. (1986) 'Sex preference, fertility, and family planning in China', *Population and Development Review* 12: 221–46.

Arnold, J.E.M. (1990) 'Tree components in farming systems', *Unasylva* 41 (106): 35–42.

Arrhenius, S. (1896) 'The influence of the carbonic acid on the air temperature of the ground', *Philosophical Magazine* (Series 5) 41: 237–76.

Attfield, R. (1983) Christian attitudes to nature. *Journal of the History of Ideas* 44: 369–86.

Atwater, W.O. 1895. *Food as Related to Life and Survival*, Chicago, IL: C.H. Kerr & Co.

Ausubel, J.H. (1991a) 'Stormy weather?', *American Scientist* 79: 292.

Ausubel, J.H. (1991b) 'A second look at the impacts of climate change', *American Scientists* 79: 210–21.

Baird, G. *et al.* (1984) *Energy Performance of Buildings*, Boca Raton, FL: CRC Press.

Bajura, R.A. and Webb, H.A. (1991) 'The marriage of gas turbines and coal', *Mechanical Engineering* 113 (9): 58–62.

Banister, J. (1988) *China's Changing Population*, Stanford, CA: Stanford University Press.

217

REFERENCES

Barbier, E.B. (1987) 'The concept of sustainable economic development', *Environmental Conservation* 14: 101–10.

Barnes, R.D. (1989) 'Diversity of organisms: how much do we know?' *American Zoologist* 29: 1075–84.

Barney, G.O. (ed.) (1980) *The Global 2000 Report to the President*, Washington, DC: US Government Printing Office.

Bassuk, E.L. (1991) 'Homeless families', *Scientific American*, 265 (6): 66–74.

Bates, H.W. (1863) *The Naturalist on the River Amazon*, London: Murray.

Batty, J.C. and Keller, J. (1980) 'Energy requirements for irrigation', in D. Pimentel (ed.) *Handbook of Energy Utilization in Agriculture*, Boca Raton, FL: CRC Press, 35–44.

Baxter, V. and Fairchild, P. (1990) 'Examining substitutes for ozone-depleting chemicals', *ORNL Review* 23 (3): 18–26.

Bazzaz, F.A. (1990) 'The response of natural ecosystems to the rising global CO_2 level', *Annual Review of Ecology and Systematics* 21: 167–96.

Berdyaev, N. (1936) *The Meaning of History*, New York, NY: Scribner.

Berry, B.J.L. (1988) 'Review of *Theory of Population and Economic Growth*', *Economic Development and Cultural Change* 37: 223.

Bethel, T. (1976) 'Darwin's mistake', *Harper's* 252: 70–5.

Bezdicek, D.F. *et al.* (eds) (1984) *Organic Farming: Current Technology and its Role in a Sustainable Agriculture*, Madison, WI: American Society of Agronomy.

Bhattacharya, A.K. (1986) 'Protein-energy malnutrition (kwashiorkor-marasmus syndrome): terminology, classification, and evolution', *World Review of Nutrition and Dietetics* 47: 80–133.

Binswanger, H.P. (1991) 'Brazilian policies that encourage deforestation in the Amazon', *World Development* 19: 821–9.

Blackburn, F. (1984) *Sugar Cane*, New York, NY: John Wiley.

Blaikie, P. and Brookfield, H. (eds) (1987) *Land Degradation and Society*, London: Methuen.

Boden, T.A. *et al.* (1990) *Trends '90*, Oak Ridge, TN: Carbon Dioxide Information Analysis Center.

Bohm, J. (1982) *Electrostatic Precipitators*, Amsterdam: Elsevier.

Bolin, B. and Cook, R.B. (eds) (1983) *The Major Biogeochemical Cycles and their Interactions*, New York, NY: John Wiley.

Borges, J.L. (1957) *Obras completas III*, Buenos Aires: Emece Edition.

Bormann, H. (1990) 'Air pollution and temperate forests: creeping degradation', in G.M. Woodwell (ed.) *The Earth in Transition*, Cambridge: Cambridge University Press, 25–44.

Borrini, G. and Margen, S. (1985) *Human Energetics*, Ottawa: IDRC.

Boulding, K.E. (1973) 'The shadow of the stationary state', *Daedalus* Fall 1973: 89–101.

Boulding, K.E. (1974) 'The social system and the energy crisis', *Science* 184: 255–7.

Bowsher, C.A. (1989) 'Enormous modernization and cleanup problems in the nuclear weapons complex', statement before the Subcommittee on Transportation and Hazardous Materials, Committee on Energy and Commerce, House of Representatives, Washington, DC.

BP (British Petroleum) (1991) *BP Statistical Review of World Energy*, London: BP.

Bramwell, A. (1989) *Ecology in the 20th Century: A History*, New Haven, CT: Yale University Press.

Brimblecombe, P. (1987) *The Big Smoke*, London: Routledge.

Broadus, J.M. (1989) 'Impacts of future sea level rise', in R.S. DeFries and T.F.

218

Malone (eds) *Global Change and Our Common Future*, National Academy Press, Washington, DC, 125–38.

Brooks, D.R. and Wiley, E.O. (1986) *Evolution as Entropy*, Chicago, IL: University of Chicago Press.

Brown, L.R. (1976) *In the Human Interest*, New York, NY: W.W. Norton.

Brown, L.R. (1989) 'Global ecology at the brink', *Challenge* 32 (2): 14–22.

Brown, S. and Lugo, A.E. (1990) 'Tropical secondary forests', *Journal of Tropical Ecology* 6: 1–32.

Bryant, E.A. (1987) 'CO_2-warming, rising sea-level and retreating coasts: review and critique', *Australian Geographer* 18: 101–13.

Buber, M. (1937) *I and Thou*, New York, NY: Sribner.

Burchell, R.W. and Listokin, D. (1982) *Energy and Land Use*, Piscataway, NJ: Center for Urban Policy Research.

Butzer, K.W. (1980) 'Civilizations: organisms or systems?' *American Scientist* 68: 517–23.

Caldwell, J.C. and Caldwell, P. (1987) 'The cultural context of high fertility in sub-Saharan Africa', *Population and Development Review* 13: 409–37.

Carr-Saunders, A. (1936) *World Population: Past Growth and Present Trends*, Oxford: Oxford University Press.

Carroll, L. (1870) *Through the Looking Glass*, London: Macmillan.

Carroll, L. (1875) *The Hunting of the Snark*, London: Macmillan.

Carson, R.L. (1962) *Silent Spring*, Boston, MA: Houghton Mifflin.

Cassidy, D.C. (1992) 'Heisenberg, uncertainty and the quantum revolution', *Scientific American* 266 (5): 100–12.

CAST (Council for Agricultural Science and Technology) (1976) *Effect of Increased Nitrogen Fixation on Stratospheric Ozone*, Ames, IO: CAST.

Chardin, P.T. de. (1966) *Man's Place in Nature*, New York, NY: Harper & Row.

Charlson, R.J. *et al.* (1992) Climate forcing by anthropogenic aerosols. *Science* 255: 423–9.

Cherrett, J.M. (1988) Ecological concepts – a survey of the view of the members of the British Ecological Society. *Biologist* 35: 64–6.

Chileske, J.H. (1988) *Nothing Wrong with Africa Except . . .*, New Delhi: Vikas Publishing.

Chowdhury, M.K. and Bairagi, R. (1990) 'Son preference and fertility in Bangladesh', *Population and Development Review* 16: 749–57.

Clark, E.L. (1986) Cogeneration – efficient energy source. *Annual Review of Energy* 11: 275–94.

Clark, W.C. and Munn, R.E. (eds) (1986) *Sustainable Development of the Biosphere*, Cambridge: Cambridge University Press.

Cochran, T.B. and Norris, R.S. (1991) 'A first look at the Soviet bomb complex', *Bulletin of the Atomic Scientists* 47 (4): 25–31.

Cody, B. (1988) *Debt-for-Nature Swaps in Developing Countries*, Washington, DC: Congressional Research Service.

Cohen, J.E. (1971) 'Mathematics as metaphor', *Science* 172: 674–5.

Colwell, R.K. (1985) 'The evolution of ecology', *American Zoologist* 25: 771–7.

Committee on Science, Engineering, and Public Policy (1991) *Policy Implications of Greenhouse Warming*, Washington, DC: National Academy Press.

Commoner, B. (1971) *The Closing Circle*, New York, NY: Knopf.

Connell, J.H. and Lowman, M.D. (1989) 'Low-diversity tropical rainforests: some possible mechanisms for their existence', *The American Naturalist* 134: 88–119.

Connell, J.H. and Sousa, W.P. (1983) 'On the evidence needed to judge ecological stability or persistence', *The American Naturalist* 121: 789–824.

REFERENCES

Connor, E.F. and Simberloff, D. (1986) 'Competition, scientific method, and null models in ecology', *American Scientist* 74: 155–62.

Conservation Technology Information Center (1989) *National Survey of Conservation Tillage Practices*, West Lafayette, IN: Conservation Technology Information Center.

Cosmovici, C.B. (ed.) (1973) *Supernovae and Supernova Remnants*, Dordrecht: Reidel.

Costanza, R. (ed.) (1991) *Ecological Economics*, New York, NY: Columbia University Press.

Crocker, T.D. (1966) 'The structuring of atmospheric pollution control systems', in H. Wolozin, (ed.), *The Economics of Air Pollution*, New York, NY: Norton, 61–86.

Crosson, P. (1985) 'National costs of erosion effects on productivity', *Erosion and Soil Productivity*, St. Joseph, MI: American Society of Agricultural Engineers, 254–65.

Cullen, R. (1986) 'Himalayan mountaineering expedition garbage', *Environmental Conservation* 13: 293–7.

Curry, R.R. (1977) 'Reinhabiting the Earth: life support and the future primitive', in J. Cairns, *et al.*, (eds) *Recovery and Restoration of Damaged Ecosystems*, Charlottesville, VA: University Press of Virginia, 1–23.

Daigger, L. (1974) 'Scientific analysis compares cost of fertility', *Upbeet* 62: 10.

Daly, H. (1991) 'Sustainable growth: a bad oxymoron', *Orion* Spring1991: 1.

Daly, H. and Cobb, J.B. (1989) *For the Common Good: Redirecting the Economy Toward Community, the Environment, and a Sustainable Future*, Boston, MA: Beacon Press.

Darwin, C. (1859) *On the Origin of Species*, London: John Murray.

Day, L.H. (1971) 'Concerning the optimum level of population', in S.F. Singer (ed.) *Is There an Optimum Level of Population?*, New York, NY: McGraw-Hill, 278.

De Gregori, T.R. (1987) 'Resources are not; they become: an institutional theory', *Journal of Economic Issues* 21: 1241–63.

DeFries, R.S. and Malone, T.F. (eds) (1989) *Global Change and our Common Future*, Washington, DC: National Academy Press.

Demeny, P. (1989) 'Demography and the limits to growth', in M.S.Teitelbaum and J.M. Winter (eds) *Population and Resources in Western Intellectual Traditions*, New York, NY: The Population Council, 213–44.

Demeny, P. (1990) 'Long-term population change', in B.L. Turner II *et al.* (eds) *The Earth as Transformed by Human Action*, Cambridge: Cambridge University Press, 25–39.

Denning, P.J. (1990) 'Modeling reality', *American Scientist* 78: 495–8.

Desurvire, E. (1992) 'Lightwave communications: the fifth generation', *Scientific America* 266 (1): 114–21.

Diamond, J. (1988) 'The last first encounters' *Natural History* 97 (8): 28–31.

Dixon, R.O.D. and Wheeler, C.T. (1986) *Nitrogen Fixation in Plants*, London: Chapman and Hall.

Domrös, M. and Peng, G. (1988) *The Climate of China*, Berlin: Springer-Verlag.

Doorenbos, J., *et al.* (1979) *Yield Response to Water*, Rome: FAO.

Dourojeanni, M.J. (1989) 'View from the South', in R.S. DeFries and T.F. Malone (eds) op. cit., 198–203.

Dudal, R. (1987) 'Land resources for plant production', in D.J.McLaren and B.K. Skinner (eds) *Resources and World Development*, Chichester: John Wiley, 659–70.

Eamus, D. and Jarvis, P.G. (1989) 'The direct effects of increase in the global atmospheric CO_2 concentration on natural and commercial temperate trees and forests', *Advances in Ecological Research* 19: 1–41.

Eckholm. E. and Brown, L.R. (1977) 'Spreading deserts – the hand of man', *Worldwatch Paper Nr. 13*, Washington, DC: Worldwatch.

Eckholm. E.P. (1976) *Losing Ground*, New York, NY: W.W. Norton.

Eden, M.J. (1990) *Ecology and Land Management in Amazonia*, London: Belhaven Press.

Edmundson, W.C. and Sukhatme, P.V. (1990) 'Food and work: poverty and hunger?' *Economic Development and Cultural Change* 38: 263–80.

Edwards, C.A. *et al.*, (eds) (1990) *Sustainable Agricultural Systems*, Ankeny, IO: Soil and Water Conservation Society.

Ehrenfeld, D. (1986) 'Why put a value on biodiversity', in E.O.Wilson, (ed.), *Biodiversity*, Washington, DC: National Academy Press, 212–16.

Ehrlich, P.R. (1968) *The Population Bomb*, New York, NY: Ballantine.

Ehrlich, P.R. and Ehrlich, A. (1981) *Extinction*, New York, NY: Random House.

Ehrlich, P.R. and Ehrlich, A.H. (1990) *The Population Explosion*, New York, NY: Simon & Schuster.

Ehrlich, P.R. and Holdren, J. (eds) (1988) *The Cassandra Conference: Resources and the Human Predicament*, College Station, TX: Texas A&M University Press.

Ehrlich, P.R. and Wilson, E.O. (1991) 'Biodiversity studies: science and policy', *Science* 253: 758–62.

Ehrlich, P.R. *et al.* (1977) *Ecoscience*, San Francisco, CA: W.H. Freeman

Eisner, R. (1991) 'The real rate of U.S. national saving, *Review of Income and Wealth* 37: 15–32.

El-Sayed, S.Z. (1988) 'Fragile life under the ozone hole', *Natural History* 97 (10): 73–80.

Ellis, D. (1989) *Environments at Risk*, Berlin: Springer-Verlag.

Ellul, J. (1964) *The Technological Society*, New York, NY: A.A.Knopf.

El Serafy, S. (1991) 'The environment as capital', in R. Costanza, op. cit., 168–75.

Emerson, R.W. (1836) *Nature*, Boston, MA: J. Munroe & Co.

Emerson, R.W. (1841) *The Method of Nature*, Boston, MA: Samuel G.Simpkins.

Emerson, R.W. (1860) *The Conduct of Life*, Boston. MA: Ticknor and Fields.

Endler, J.A. and McLellan, T. (1988) 'The processes of evolution: toward a newer synthesis', *Annual Review of Ecology and Systematics* 19: 395–421.

Engelstad, O.P. (ed.) (1985) *Fertilizer Technology and Use*, Madison, WI: Soil Science Society of America.

Enoch, H.Z. and Kimball, B.A. (eds) (1986) *Carbon Dioxide Enrichment of Greenhouse Crops: Status and Carbon Dioxide Sources*, Boca Raton, FL: CRC Press.

EPRI (Electric Power Research Institute) (1981) *EPRI Sulfate Regional Experiment: Results and Implications*, Palo Alto, CA: EPRI.

Epstein, J.M. and Gupta, R. (1990) *Controlling the Greenhouse Effect*, Washington, DC: The Brookings Institution.

Erwin, T.L. (1986) 'The tropical forest canopy', in E.O. Wilson (ed.) op. cit., 123–9.

Evans, G.C. (1976) 'A sack of uncut diamonds: the study of ecosystems and the future resources of mankind', *Journal of Applied Ecology* 13: 1–39.

Falkenmark, M. (1986) 'Fresh water – time for a modified approach', *Ambio* 15: 192–200.

Falkenmark, M. (1989) 'The massive water scarcity now threatening Africa – why isn't it being addressed?' *Ambio* 18: 112–18.

FAO (Food and Agriculture Organization) (1980) *A Global Reconnaisance Survey of the Fuelwood Supply/Requirement Situation*, Rome: FAO.

REFERENCES

FAO (1981) *Agriculture: Toward 2000*, Rome: FAO.

FAO (1990) *Fertilizer Yearbook*, Rome: FAO.

FAO (1991) *Production Yearbook*, Rome: FAO

FAO/WHO/UNU (Food and Agriculture Organization/World Health Organization/ United Nations University) Expert Consultation (1985) *Energy and Protein Requirements*, Geneva: WHO.

Farman, J.C. *et al.* (1985) 'Large losses of total ozone in Antarctica reveal seasonal ClO_x/NO_x interaction', *Nature* 315: 201–10.

Fearnside, P.M. (1989) 'Extractive reserves in Brazilian Amazonia', *BioScience* 39: 387–93.

Fearnside, P.M. (1990) 'Deforestation in Brazilian Amazonia', in G.M. Woodwell (ed.) *The Earth in Transition*, Cambridge: Cambridge University Press, 211–38.

Ferguson, E.E. (ed.) (1991) *Climate Monitoring and Diagnostics Laboratory No. 19 Summary Report 1990*, Boulder, CO: NOAA.

Ferris, B.G. (1969) 'Chronic low-level air pollution', *Environmental Research* 2: 79–87.

Feshbach, M. and Friendly, A. (1992) *Ecocide in the USSR*, New York, NY: Basic Books.

Feyerabend, P. (1975) *Against Method*, London: NLB.

Feynman, R.P. (1963) *The Feynman Lectures on Physics*, Reading, MA: Addison-Wesley.

Fitjer, H. (1990) 'Small hydro development in Ethiopia', *Water Power & Dam Construction* 42 (10): 33–8.

Forrester, J.W. (1971) *World Dynamics*, Boston, MA: Wright-Allen Press.

Fox, R.L. and Yost, R.S. (1980) 'Estimating global fertilizer requirements: some tentative results', in R.P. Sheldon *et al.*, (eds) *Fertilizer Raw Material Resources, Needs and Commerce in Asia and the Pacific*, Honolulu, HI: East–West Center, 211–24.

Frederick, K.D. (1988) 'Irrigation under stress', *Resources* Spring 1988: 1–4.

Frederick, K.D. (1991) 'The disappearing Aral Sea', *Resources* Winter 1991: 11–14.

Friedman, H. (1986) *Sun and Earth*, New York, NY: Scientific American Library.

Friis-Christensen, E. and Lassen, K. (1991) *Science* 254: 698–700.

Frohlich, C. (1987) 'Variability of the solar "constant" on timescales of minutes to years', *Journal of Geophysical Research* 4: 277–92.

Frosch, R.A. and Gallopoulos, N.E. (1989) 'Strategies for manufacturing', *Scientific American* 261 (3): 143–52.

GAO (General Accounting Office) (1990) *Foreign Assistance International Resource Flows and Development Assistance to Developing Countries*, Washington, DC: GAO.

GAO (General Accounting Office) (1991a) *Electricity Supply*, Washington, DC: GAO.

GAO (General Accounting Office) (1991b) *Persian Gulf: Allied Burden Sharing Efforts*, Washington, DC: GAO.

Gaston, K.J. (1991) 'The magnitude of global insect species richness', *Conservation Biology* 5: 283–96.

Gebhardt, M.R. *et al.* (1985) Conservation tillage. *Science* 230: 625–30.

Getis, J. (1991) *You Can Make a Difference*, Dubuque, IO: W.C. Brown.

Gilland, B. (1985) 'Cereal yields in theory and practice', *Outlook on Agriculture* 14: 56–60.

Gillis, A.M. (1992) 'Shrinking the trash heap', *BioScience* 42: 90–3.

Gleick, J. (1987) *Chaos: Making a New Science*, New York, NY: Viking Press.

Goeller, H.E. and Zucker, A. (1984) 'Infinite resources: the ultimate strategy',

Science 223: 456–62.

Golerman, H.L., (ed.) (1986) *Denitrification in the Nitrogen Cycle*, New York, NY: Plenum Press.

Goodland, R.J.A. *et al.* (1991) 'Tropical moist forest management: the urgency of transition to sustainability', in R.Costanza, op. cit., 486–515.

Gottlieb, R. (1991) *Thirst for Growth*, Tucson, AZ: University of Arizona Press.

Grant, J.P. (1990) *The State of the World's Children 1991*, Oxford: Oxford University Press.

Grant, P.R. (1991) 'Natural selection and Darwin's finches', *Scientific American* 265 (4): 82–7.

Greene, J.C. (1989) 'Introductory conversation', in J.R. Morre, (ed.) *History, Humanity and Evolution*, Cambridge: Cambridge University Press.

Haagen-Smit, A.J. (1970) 'A lesson from the smog capital of the world', *Proceedings of the National Academy of Sciences* 67: 887–97.

Hacking. I. (ed.) (1981) *Scientific Revolutions*, Oxford: Oxford University Press.

Hafele, W. (1981) *Energy in a Finite World: A Global Systems Analysis*, Laxenburg: International Institute for Applied Systems Analyst.

Hall, J.V. *et al.* (1992) 'Valuing the health benefits of clean air', *Science* 255: 812–16.

Hall, P. (1984) *The World Cities*, New York, NY: St. Martin's Press.

Hanna, J.M. and Brown, D.E. (1983) 'Human heat tolerance: an anthropological perspective', *Annual Review of Anthropology* 12: 259–84.

Hansen, J. and Lebedeff, S. (1988) 'Global surface air temperatures: Update through 1987', *Geophysical Research Letters* 15: 323–6.

Hanson, H. *et al.* (1982) *Wheat in the Third World*, Boulder, CO: Westview Press.

Hardin, G. (1974) 'Living on a lifeboat', *BioScience* 24: 561–8.

Harlin, J.M. and Berardi, G.M. (eds) (1987) *Agricultural Soil Loss*, Boulder, CO: Westview Press.

Harris, C. (1991) 'Frontrunners of China's agriculture', *IDRC Reports* 1991 (July): 16–17.

Harrison, M. (1988) 'Resource mobilization for World War II: the USA, UK, USSR, and Germany, 1938–1945', *Economic History Review* 41: 171–92.

Hartenstein, R. (1986) 'Earthworm biotechnology and global biogeochemistry', *Advances in Ecological Research* 15: 379–409.

Harvell, M.A. (1986) *Nuclear Winter*, New York, NY: Springer-Verlag.

Haugerud, A. and Collinson, M.P. (1991) *Plants, Genes and People: Improving the Relevance of Plant Breeding*, London: International Institute for Environment and Development.

Heilbroner, R.L. (1974) *An Inquiry into the Human Prospect*, New York, NY: W.W. Norton.

Hellden, U. (1991) 'Desertification – time for an assessment?' *Ambio* 20: 372–83.

Hellriegel, H. and Wilfarth, H. (1888) *Untersuchungen über die Stickstoffnährung der Graminen und Leguminosen*, Berlin: Kayssler & Co.

Henderson-Sellers, A. and Gornitz, V. (1984) 'Possible climatic impacts of land cover transformation, with particular emphasis on tropical deforestation', *Climatic Change* 6: 231–57.

Hendry, J. (1990) 'Humidity, hygiene, or ritual care: some thoughts on wrapping as a social phenomenon', in E. Ben-Ari *et al.* (eds) *Unwrapping Japan*, Honolulu, HI: University of Hawaii Press, 18–35.

Heng, L. and Shapiro, J. (1986) *After the Nightmare*, New York, NY: A.A. Knopf.

Herbertson, J.H. (1913) 'The higher units', *Scientia* 14: 199–212.

Herskowitz, A. and Salerni, E. (1989) 'Municipal solid waste incineration in Japan', *Environmental Impact Assessment Review* 9: 257–78.

Hickman, B.G. (ed.) (1983) *Global International Economic Models*, New York, NY: North-Holland.

Higgitt, D.L. (1991) 'Soil erosion and soil problems', *Progress in Physical Geography* 15: 91–100.

Higgs, R.L. *et al.* (1990) 'Crop rotations: sustainable and profitable', *Journal of Soil and Water Conservation* 45: 68–70.

Hinrichsen, D. (1986) 'Multiple pollutants and forest decline', *Ambio* 15: 258–65.

Hofstetter, R.H. (1983) 'Wetlands in the United States', in A.J.P. Gore (ed.) *Mires: Swamps, Bogs, Fen and Moor*, Amsterdam: Elsevier, 201–44.

Holdermann, K. (1953) *Im Banne der Chemie: Carl Bosch, Leben und Werke*, Düsseldorf: Econ-Verlag.

Holdren, J.P. (1990) 'Energy in transition', *Scientific American* 263 (3): 156–63.

Holland, H.D. (1985) 'The oxygen content of the atmosphere', *Earth and Mineral Sciences* 55 (2): 14–17.

Houghton, J.T. *et al.* (eds) (1990) *Climate Change: The IPCC Scientific Assessment*, Cambridge: Cambridge University Press.

Houghton, R.A. and Skole, D.L. (1990) 'Carbon', in B.L. Turner II *et al.* (eds), op. cit., 387–408.

Hu, S.D. (1983) *Handbook of Industrial Energy Conservation*, New York, NY: Van Nostrand Reinhold.

Huang, R. (1988) 'Development of groundwater for agriculture in the lower Yellow River alluvial basin', in G.T. O'Mara, (ed.) *Efficiency in Irrigation*, Washington, DC: World Bank, 80–4.

Hubbard, H.M. (1989) 'Photovoltaics today and tomorrow', *Science* 244: 297–304.

Hughes, T.P. (1983) *Networks of Power*, Baltimore, MD: Johns Hopkins University Press.

von Humboldt, A. (1849) *Ansichten der Natur*, Stuttgart: Cotta.

Humphrey, W.S. and Stanislaw, J. (1979) 'Economic growth and energy consumption in the UK, 1700–1975', *Energy Policy* 7: 29–42.

Hutchinson, T.C. and Meema, K.M. (eds) (1987) *Lead, Mercury, Cadmium and Arsenic in the Environment*, New York, NY: John Wiley.

INSEE (Institut national de la statistique et des études économiques) (1955–91) *Annuaire Statistique de la France*, Paris: INSEE.

International Commission on Large Dams (1978) *Environmental Effects of Large Dams*, New York, NY: American Society of Civil Engineers.

IUCN/UNEP/WWF (International Union for the Conservation of Nature/United Nations Environment Porgramme/World Wildlife Fund) (1991) *Caring for the Earth: A Strategy for Sustainable Living*, Gland: IUCN.

Jacquard, A. (1985) *Endangered by Science?* New York, NY: Columbia University Press.

Jee, K.K. (1988) 'Environmental improvement in Singapore', *Ambio* 17: 233–7.

Jensen, W.G. (1970) *Energy and the Economy of Nations*, Henley-on-Thames: G.T. Foulis.

John Paul II (1991) *Centessimus Annus*, Rome: Vatican.

Johnels, A. (1983) 'Conference on environmental research and management priorities', *Ambio* 12 (2): 58–120.

Johnson, C. *et al.* (1992) 'Impact of aircraft and surface emissions of nitrogen oxides on tropospheric ozone and global warming', *Nature* 355: 69–71.

Jones, P.D. *et al.* (1986) 'Global temperature variations between 1861 and 1984', *Nature* 322: 430–4.

Kammerer, J.C. (1982) 'Estimated demand of water for different purposes', in *Water for Human Consumption*, International Water Resources Association, Dublin:

Tycooly Publishing, 141–72.

Karl, T.R. *et al.* (1991) 'Global warming – evidence for asymmetric diurnal temperature change', *Geophysical Research Letters* 18: 2253–6.

Karlovsky, J. (1981) 'Cycling of nutrients and their utilization by plants in agricultural ecosystems', *Agro-Ecosystems* 2: 127–44.

Kasting, J.P. (1989) 'Long-term stability of the Earth's climate', *Palaeogeography, Palaeoclimatology, Palaeoecology* 75: 83–95.

Katz, S. (1984) *Classic Plastics: from Bakelite to High-tech*, London: Thomas and Hudson.

Keay, R.W.J. (1990) 'Presidential Address 1990: a sense of perspective in the tropical rain forest', *Biologist* 37: 73–8.

Kerr, R.A. (1990a) 'Climatologists debate how to model the world', *Science* 250: 1082–3.

Kerr, R.A. (1990b) 'Another deep Antarctic ozone hole', *Science* 250: 370.

Kerr, R.A. (1991) 'Greenhouse bandwagon rolls on', *Science* 253: 845.

Keyfitz, N. (1989) 'The growing human population', *Scientific American* 261(3): 119–26.

Kimball, B.A. (1983) 'Carbon dioxide and agricultural yield: an assemblage and analysis of 430 prior observations', *Agronomy Journal* 75: 779–88.

Klein, R.M. and Perkins, T.D. (1987) 'Cascades of causes and effects of forest decline', *Ambio* 16: 86-93.

Klein, T.F.D. (1980) 'Renaissance man', in I. Asimov *et al.*, (eds) *Microcosmic Tales*, New York, NY: Taplinger Publishing, 59–64.

Komarov, B. (1980) *The Destruction of Nature in the Soviet Union*, Armonk, NY: M.E. Sharpe.

Korzybski, A. (1933) *Science and Sanity*, Lakeville, CT: The International Non-Aristotelian Library.

Krug, E.C. and Frink, C.R. (1983) 'Acid rain on acid soil: a new perspective', *Science* 221: 520–5.

Krupnick, A.J. and Portney, P.R. (1991) 'Controlling urban air pollution: a benefit-cost assessment', *Science* 252: 522–8.

Lacey, R. (1986) *Ford, the Men and the Machine*, Toronto, ON: McClelland and Stewart.

Lal, R. (1990) *Soil Erosion in the Tropics: Principles and Management*, New York, NY: McGraw-Hill.

Landsberg, H.H. (1988) '*Resources for Freedom* in retrospect', *Resources* 90: 7–12.

Lanly, J.-P. (1982) *Tropical Forest Resources*, Rome: FAO.

Larson, E.D. *et al.* (1986) 'Beyond the era of materials', *Scientific American* 254: 34–41.

Larson, W.E. *et al.* (1983) 'The threat of soil erosion to long-term crop production', *Science* 219: 458–65.

Laudan, L. (1984) 'Explaining the success of science: beyond epistemic realism and relativism', in J.T. Cushing *et al.* (eds) *Science and Reality*, Notre Dame, IN: University of Notre Dame Press, 83–104.

Lee, L.K. (1984) 'Land use and soil loss: a 1982 update', *Journal of Soil and Water Conservation* 39: 226–8.

Lee, L.K. (1990) 'The dynamics of declining soil erosion rates', *Journal of Soil and Water Conservation* 45: 622–4

Lele, S.M. (1991) 'Sustainable development: a critical review', *World Development* 19: 607–21.

Lemon, E.R., (ed.) (1983) CO_2 *and Plants*, Boulder, CO: Westview Press.

Leontief, W. (1982) 'Academic economics', *Science* 217: 104–7.

225

Lifton, R.J. (1985) 'Toward a nuclear-age ethos', *Bulletin of the Atomic Scientists* 41 (7): 172.

Likens, G.E. and Butler, T.J. (1981) 'Recent acidification of precipitation in North America', *Atmospheric Environment* 15: 1103–9.

Lindeman, R.L. (1942) 'The trophic-dynamic aspects of ecology', *Ecology* 23: 399–418.

Lindheim, R. and Syme, S.L. (1983) 'Environments, people and health', *Annual Review of Public Health* 4: 335–59.

Lindzen, R.S. (1990) 'Some remarks on global warming', *Environmental Science and Technology* 24: 424–6.

Lipfert, F.W. *et al.* (1991) 'Air pollution benefit-cost assessment', *Science* 253: 606.

Lippmann, M. (1991) 'Health effects of tropospheric ozone', *Environmental Science & Technology* 25: 1954–61.

Longhurst, J.W., (ed.) (1991) *Acid Deposition Origin, Impacts and Abatement Strategies*, Berlin: Springer-Verlag.

Lorenz, E.N. (1979) 'Predictability: does the flap of a butterfly's wings in Brazil set off a tornado in Texas?' Paper presented at the American Association for Advancement of Science Annual Meeting, Houston, 3–8 January 1979.

Lotka, A.J. (1925) *Elements of Mathematical Biology*, Baltimore, MD: Williams and Wilkins.

Lovelock, J.E. (1979) *Gaia*, Oxford: Oxford University Press.

Lowi, M. (1992). 'West Bank water resources and the resolution of conflict in the Middle East', Boston, MA: Occasional paper of the American Academy of Arts and Sciences.

Lugo, A.E. (1986) 'Estimating reductions in the diversity of tropical forest species', in E.O. Wilson (ed.) op. cit., 58–70.

L'vovich, M.I. *et al.* (1990) 'Use and transformation of terrestrial water systems', in B.L. Turner II *et al.* (eds) op. cit., 235–52.

Lynam, J.K. and Herdt, R.W. (1988) 'Sense and sustainability as an objective in international agricultural research', Paper prepared for CIP-Rockefeller Foundation Conference on 'Farmers and Food Systems', Lima, Peru, September 26–30, 1988.

MacArthur, R.H. and Wilson, E.O. (1963) 'An equilibrium theory of island biogeography', *Evolution* 17: 373–87.

McCloskey, J.M. and Spalding, H. (1989) 'A reconnaissance-level inventory of the amount of wilderness remaining in the world', *Ambio* 18: 221–7.

McKean, R.N. (1973) 'Growth vs. no growth: an evaluation', *Daedalus*, Fall 1973: 217.

McKenny, D.J. *et al.* (1978) 'Rates of N_2O evolution from N-fertilized soils', *Geophysical Research Letters* 5: 530–2.

McKnight, T.L. (1990) 'Irrigation technology: a photo-essay', *Focus* 40: 1–6.

McLaren, D.J. and Skinner, B.J. (eds) (1987) *Resources and World Development*. Chichester: John Wiley.

Macriss, R.A. (1983) 'Space heating by gas and oil', *Annual Review of Energy* 8: 247–67.

Mahar, D.J. (1989) 'Deforestation in Brazil's Amazon region: magnitude, rate, and causes', in G. Schramm and J.J. Warford, (eds) *Environmental Management and Economic Development*, Baltimore, MD: The Johns Hopkins University Press, 87–116.

Mahlman, J.D. (1989) 'Mathematical modeling of greenhouse warming: how much do we know?' in DeFries and Malone, (eds), op. cit., 62–72.

Malone, T.F. and Roederer, J.G. (eds) (1985) *Global Change*, Cambridge:

International Council of Scientific Unions Press and Cambridge University Press.

Makarewicz, J.C. and Bertram, P. (1991) 'Evidence for the restoration of the Lake Erie ecosystem', *BioScience* 41: 216–23.

Malthus, R. (1798) *An Essay on the Principle of Population and its Effects on Future Improvement of Society*, London: J. Johnson.

Malthus, R. (1803) *An Essay on the Principle of Population and its Effects on Future Improvement of Society*, 2nd edn, London: J. Johnson.

Mann, C.C. (1991) 'Extinction: are ecologists crying wolf?', *Science* 253: 726–8.

Mann, C.C. and Plummer, M.L. (1992) 'The butterfly problem', *The Atlantic Monthly* 219 (1): 47–70.

Manzer, L.E. (1990) 'The CFC-ozone issue: progress on the development of alternatives to CFCs', *Science* 249: 31–5.

Marchetti, C. (1977) 'On geoengineering and the CO_2 problem', *Climatic Change* 1: 59–68.

Marchetti, C. and Nakicenovic, N. (1979) *The Dynamics of Energy Systems and the Logistic Substitution Model*, Laxenburg: International Institute for Applied Systems Analysis.

Margulis, L. and Guerrero, R. (1989) 'From planetary atmospheres to microbial communities: a stroll through space and time', in D.B. Botkin (ed.) *Changing the Global Environment*, New York, NY: Academic Press, 51–67.

Marsh, G.P. (1864) *Man and Nature*. New York, NY: C. Sribner.

Marshall, L. (ed.) (1913) *The Story of Polar Conquest*, Philadelphia, PA: The John C. Winston Company.

Martinez–Alier, J. (1987) *Ecological Economics*, Oxford: Basil Blackwell.

Mather, A.S. (1990) *Global Forest Resources*, London: Bellhaven Press.

Mathews, J.T. (ed.) (1991) *Preserving the Global Environment: The Challenge of Shared Leadership*, New York, NY: W.W. Norton.

Matthews, R.M. (1989) *Legumes*, New York, NY: Marcel Dekker.

Meadows, D.H. *et al.* (1972) *The Limits to Growth*, New York, NY: Universe Books.

Meier, R.L. and Abdul Quium, A.S.M. (1991) 'A sustainable state for urban life in poor societies: Bangladesh', *Futures* 23: 128–45.

Melko, M. (1971) 'Is western civilization unique?' *Civilisations* 21: 39.

Merriam, M.F. (1977) 'Wind energy for human needs', *Technology Review* 79 (3): 29–39.

Metz, K.J. (1991) 'A reassessment of the causes and severity of Nepal's environmental crisis', *World Development* 29: 805–20.

Milliman, J.D. *et al.* (1989) 'Environmental and economic implications of rising sea level and subsiding deltas: the Nile and Bengal examples', *Ambio* 18: 340–5.

Ministerie van Volkshuisvesting, Ruimtelijke Ordening en Milieubeheer, Ministerie van Lanbouw en Visserij and Ministerie van Verkeer en Waterstaat (1985) *Indicatief Meerjaren Programma Milieubeheer 1986–1990*, den Haag: Tweede Kamer.

Mitchell, B.R. (1975) *European Historical Statistics, 1750–1970*, New York, NY: Columbia University Press.

Mitchell, J.F.B. *et al.* (1989) 'CO_2 and climate: a missing feedback? *Nature* 341: 132–4.

Molina, M.J. and Rowland, F.S. (1974) 'Stratospheric sink for chlorofluoromethanes: chlorine atom-catalysed destruction of ozone', *Nature* 249: 810–12.

Mooney, H.A. *et al.* (1991) 'Predicting ecosystem responses to elevated CO_2 concentrations', *BioScience* 41: 96–104.

Morrisette, P.M. and Plantinga, A.J. (1991) 'The global warming issue: viewpoints of different countries', *Resources* Spring 1991: 2–6.

Mossman, B.T. *et al.* (1990) 'Asbestos: scientific developments and implications for public policy', *Science* 247: 294–300.

Mudahar, M.S. and Hignett, T.P. (1987) 'Fertilizer and energy use', in Z.R. Helsel (ed.) *Energy in Plant Nutrition and Pest Control*, Amsterdam: Elsevier, 1–23.

Mungall, C. and McLaren, D.J. (1990) *Planet under Stress*, Toronto: Oxford University Press.

Munson, R.D. and Runge, C.F. (1990) *Improving Fertilizer and Chemical Efficiency through 'High Precision Farming'*, St.Paul, MN: Center for International Food and Agricultural Policy, University of Minnesota.

(MVMA) Motor Vehicle Manufacturing Association (1990) *MVMA Motor Vehicle Facts and Figures '91*, Detroit, MI: MVMA.

Myers, N. (1979) *The Sinking Ark*, Oxford, Pergamon Press.

Myers, N. (1980) 'The present status and future prospects of tropical moist forests', *Environmental Conservation* 7: 101–14.

Nader, L. and Beckerman, S. (1978) 'Energy as it relates to the quality and style of life', *Annual Review of Energy* 3: 1–28.

Nair, P.K.R. and Fernandes, E. (1984) *Agroforestry as an Alternative to Shifting Cultivation*, Rome: FAO.

NAPAP (National Acid Precipitation Assessment Program) (1990) *Annual Report*, Washington, DC: NAPAP.

NAS (National Academy of Sciences) (1980) *The Effects on Populations of Exposure to Low Levels of Ionizing Radiation: 1980*, Washington, DC: NAS.

NAS (1981) *The Health Effects of Nitrate, Nitrite and N-Nitroso Compounds*, Washington, DC: NAS.

NRC (National Research Council) (1979) *Stratospheric Ozone Depletion by Halocarbons: Chemistry and Transport*, Washington, DC: NAS.

NRC (1982) *Causes and Effects of Stratospheric Ozone Destruction: An Update*, Washington, DC: NAS.

NRC (1983) *Acid Deposition: Atmospheric Processes in Eastern North America*, Washington, DC: NAS.

NRC (1984) *Causes and Effects of Changes in Stratospheric Ozone: Update 1983*, Washington, DC: NAS.

NRC (1986) *Acid Deposition: Long-term Trends*, Washington, DC: NAS.

NRC (1989) *Alternative Agriculture*, Washington, DC: NAS.

Neal, H.A. and Schubel, J.R. (1987) *Solid Waste Management and the Environment*, Engelwood Cliffs, NJ: Prentice Hall.

Needleman, H.L. and Bellinger, D. (1991) 'The health effects of low level exposure to lead', *Annual Review of Public Health* 12: 111–40.

Nelkin, G.A. and Dellefield, R.J. (1990) 'Fluidized-bed combustion', *Mechanical Engineering* 112 (9): 58–62.

Newmark, W.D. (1987) 'A land-bridge island perspective on mammalian extinctions in western North American parks', *Nature* 325: 430–2.

Nihlgard, B. (1985) 'The ammonium hypothesis – an additional explanation to the forest dieback in Europe', *Ambio* 14: 2–8.

Nishimura, H. (1989) *How to Conquer Air Pollution*, Amsterdam: Elsevier.

Nobre, C.A. *et al.* (1991) 'Amazonian deforestation and regional climate change', *Journal of Climate* 4: 957–88.

Norton, B. (1986) 'Commodity, amenity, and morality', in E.O.Wilson (ed.) op. cit., 200–5.

Nriagu, J.O. *et al.* (1991) 'Origin of sulphur in Canadian Arctic haze from isotopic measurements', *Nature* 349: 142–4.

Nulty, P. (1991) 'Finding a payoff in environmentalism', *Fortune* 124 (9): 79–81.

Oates, W.E. (1988) 'Taxing pollution: an idea whose time has come?' *Resources* Spring 1988: 5–7.

Odum, E.P. (1971) *Fundamentals of Ecology*, Philadelphia, PA: W.B. Saunders.

Odum, H.T. (1971) *Environment, Power, and Society*, New York, NY: John Wiley.

Odum, H.T. (1983) *Systems Ecology*, New York, NY: John Wiley.

OECD (Organization for Economic Co-operation and Development) (1981) *The Costs and Benefits of Sulphure Oxide Control*, Paris: OECD.

OECD (1982) *Eutrophication of Waters: Monitoring, Assessment and Control*, Paris: OECD.

O'Leary. P.R. *et al.* (1988) 'Managing solid waste', *Scientific American* 259 (6): 36–42.

O'Mara, G.T. (1988) 'The efficient use of surface water and groundwater in irrigation: an overview of the issues', in G.T. O'Mara (ed.) *The Conjunctive Use of Surface and Groundwater Resources*, Washington, DC: World Bank, 1–17.

OPEC (Organization of Petroleum-Exporting Countries) (1991) *Facts and Figures*, Vienna: OPEC.

O'Regan, B. and Gratzel, M. (1991) 'A low-cost, high-efficiency solar cell based on dye-sensitive colloidal TiO_2 films', *Nature* 353: 737–40.

Osofsky, S.A. (1988) 'Panther diary', *Natural History* 97 (4): 50–4.

OTA (Office of Technology Assessment) (1991) *Changing by Degrees: Steps to Reduce Greenhouse Gases*, Washington, DC: OTA.

Ottar, B. (1976) 'An assessment of the OECD study on long-range transport of air pollutants', *Atmospheric Environment* 12: 445–54.

Page, D. (1988) 'Debt-for-nature swaps: fad or magic formula?' *Ambio* 17: 243–4.

Panofsky, H. (1970) 'Analyzing atmospheric behavior', *Physics Today* 23(12): 32–5.

Parker, S.P. (ed.) (1982) *Synopsis and Classification of Living Organisms*, New York, NY: McGraw-Hill.

Partridge, E. (ed.) (1981) *Responsibilities to Future Generations: Environmental Ethics*, Buffalo, NY: Prometheus Books.

Peakall, D.B. and Lovett, R.J. (1972) 'Mercury: its occurrence and effects in the ecosystem', *BioScience* 22: 20–5.

Pearce, D. 1989. 'Sustainable futures: some economic issues', in D.B. Botkin (ed.) *Changing the Global Environment*, New York, NY: Academic Press, 311–23.

Pearce, D. *et al.* (1991) *Blueprint for a Green Economy*, London: Earthscan.

Pearce, R. (1990) 'Traditional food crops in sub-Saharan Africa', *Food Policy* 15: 374–82.

Pearman, G.I. *et al.* (1986) 'Evidence of changing concentrations of atmospheric CO_2, N_2O, and CH_4 from air bubbles in Antarctic ice', *Nature* 320: 248–50.

Pechmann, J.H.K. *et al.* (1990) 'Declining amphibian populations: the problem of separating human impacts from natural fluctuations', *Science* 253: 892–5.

Pendergast, D. (1992) *The Greenhouse Effect: A New Plank in the Nuclear Power Platform*, Mississauga, ON: Atomic Energy Canada Ltd.

Perry, D.A. *et al.* (1989) 'Bootstrapping in ecosystems', *BioScience* 39: 230–7.

Peters, R.H. (1991) *A Critique for Ecology*, Cambridge: Cambridge University Press.

Petzet, G.A. (1988) 'Operators make wider use of horizontal drilling technology', *The Oil & Gas Journal* 86 (15): 15–7.

Pierce, F.J. *et al.* (1984) 'Productivity of soils in the Cornbelt: an assessment of the long-term impact of erosion', *Journal of Soil and Water Conservation* 39: 131–6.

Pigou, A.C. (1920) *The Economics of Welfare*, London: Macmillan.

Pimentel, D. *et al.* (1989) 'Low-input sustainable agriculture using ecological management practices. *Agriculture, Ecosystems and Environment* 27: 3–24.

229

Plotkin, M.J. (1986) 'The outlook for new agricultural and industrial products from the tropics', in E.O. Wilson (ed.) op. cit., 106–15.

Poincare, H. (1903) *La science et l'hypothèse*, Paris: E. Flammarion.

Precoda, N. (1991) 'Requiem for the Aral Sea', *Ambio* 20: 109–14.

Prentice, A.M. (1984) 'Adaptations to long-term low energy intake', in E. Pollitt and P. Amante (eds) *Energy Intake and Activity*, New York, NY: Alan R. Liss, 3–31.

President's Materials Policy Commission. (1952) *Resources for Freedom*, Washington, DC: US Government Printing Office.

Proctor, J. (1991) Tropical rain forests. *Progress in Physical Geography* 15: 291–303.

Prost, A. (1991) 'The family and the individual', in A. Prost and G. Vincent (eds) *A History of Private Life*, V, Cambridge, MA: Harvard University Press, 56.

Qiu, D. *et al.* (1990) 'Diffusion and innovation in the Chinese biogas program', *World Development* 18: 555–63.

Quade, E.S. (1970) *On the Limitations of Quantitative Analysis*, Santa Monica, CA: Rand Corporation.

Ramanathan, V. *et al.* (1989) 'Climate and the Earth's radiation budget', *Physics Today* 42: 22–32.

Ramberg, L. *et al.* (1987) 'Development and biological status of Lake Kariba – a man-made tropical lake', *Ambio* 16: 314–21.

Rathje, W.L. (1989) 'Rubbish!' *The Atlantic Monthly* 264 (6): 99–109.

Raup. D.M. (1981a) 'Introduction: what is a crisis?' in M.H. Nitecki (ed.) *Biotic Crises in Ecological and Evolutionary Time*, New York, NY: Academic Press, 1–12.

Raup. D.M. (1981b) 'Physical disturbance in the life of plants', in M.H. Nitecki (ed.) op. cit., 39–52.

Redclift, M. (1987) *Sustainable Development: Exploring the Contradictions*, London: Methuen.

Reij, C. (1991) *Indigenous Soil and Water Conservation in Africa*, London: International Institute for Environment and Development.

Reitz, J.R. (1985) 'Potential for energy savings in old and new auto engines' in D. Hafemeister *et al.* (eds) *Energy Sources: Conservation and Renewables*, New York, NY: American Institute of Physics.

Repetto, R., (ed.) (1985) *Global Possible*, New Haven, CT: Yale University Press.

Repetto, R. *et al.* (1991) *Accounts Overdue: Natural Resource Depletion in Costa Rica*, Washington, DC: World Resources Institute.

Reutlinger, S. and van H. Pellekaan, J. (1986) *Poverty and Hunger*, Washington, DC: World Bank.

Richards, J.F. (1990) 'Land transformation', in B.L. Turner II *et al.* (eds) op. cit., 163–78.

Richardson, G. *et al.* (1989) 'Gypsum blocks "tell a water tale"', *Journal of Soil and Water Conservation* 44: 192–5.

Richter, D.D. and Babbar, L.I. (1991) 'Soil diversity in the tropics', *Advances in Ecological Research* 21: 315–89.

Rittenhouse, R.C. (1990) 'New SO_2 limits will challenge particulate control systems', *Power Engineering* 94 (6): 18–25.

Roberts, L. (1991) 'Report nixes "Geritol" fix for global warming', *Science* 253: 1490–1.

Rocky Mountain Institute (1991) *Water-Efficient Technologies*, Old Snowmass, CO: RMI.

Rodhe, H. (1989) 'Acidification in a global perspective', *Ambio* 18: 155–9.

Rohlf, D.J. (1991) 'Six biological reasons why the Endangered Species Act doesn't work – and what to do about it', *Conservation Biology* 5: 273–82.

Rohwer, J. (1991) 'In the forest', *The Economist* Brazil Survey: 8, 7 December, 1991.

Roque, C.R. and Garcia, M.I. (1991) 'Economic inequality, environmental degradation and civil strife in the Philippines', Paper prepared for the American Academy of Arts and Sciences, Cambridge, MA.

Rosenfeld, A.H. and Hafemeister, D. (1988) 'Energy-efficient buildings', *Scientific American* 258 (4): 78–85.

Rossin, A.D. (1990) 'Experience of the US nuclear industry and requirements for a viable nuclear industry in the future', *Annual Review of Energy* 15: 153–72.

Rothenberg, J. (1989) *The Nature of Modeling*, Rand Note N3027-Defense Advanced Research Project Agenda, Santa Monica, CA: Rand Corporation.

Rousseau, D.L. (1992) Case studies in pathological science. *American Scientist* 80: 54–63.

Rowe, R. *et al.* (1987) *Santa Clara County Air Pollution Benefit Analysis*, Washington, D.C.: US Environmental Protection Agency.

Rowland, F.S. (1989) 'Chlorofluorocarbons and the depletion of stratospheric ozone', *American Scientists* 77: 36–45.

Rowland, F.S. (1991) 'Stratospheric ozone in the twenty-first century', *Environmental Science and Technology* 25: 622–8.

Royal Ministry of Foreign Affairs. (1971) *Air Pollution Across National Boundaries*, Stockholm: Royal Ministry of Foreign Affairs.

Royal Society Study Group (1984) *Nitrogen Cycle of the United Kingdom*, London: Royal Society.

Runge, C.F. *et al.* (1990) *Agricultural Competitiveness, Farm Fertilizer and Chemical Use, and Environmental Quality: A Descriptive Analysis*, Minneapolis, MN: Center for International Food and Agricultural Policy.

Sampl, F.R. and Shank, M.E. (1985) 'Aircraft turbofans: new economic and environmental benefits', *Mechanical Engineering*, 107 (9): 47–53.

Sanchez, P.A. *et al.* (1982) 'Amazon basin soils: management for continuous crop production', *Science* 216: 821–7.

Sarre, P. and Smith, P. (1991) *One World for One Earth*, London: The Open University.

Schertz, D.L. *et al.* (1985) 'Field evaluation of the effect of soil erosion on crop production', in *Erosion and Soil Productivity*, St. Joseph, MI: American Society of Agricultural Engineers, 9–17.

Schipper, L. *et al.* (1985) 'Explaining residential energy use by international bottom-up comparisons', *Annual Review of Energy* 10: 341–405.

Schlesinger, M.E. and Mitchell, J.F.B. (1987) 'Climate model simulations of the equilibrium climatic response to increased carbon dioxide', *Review of Geophysics* 25: 760–98.

Schlesinger, W.H. *et al.* (1990) 'Biological feedbacks in global desertification', *Science* 247: 1043–8.

Schlosstein, S. (1991) 'The giant Nippon monster myth', *The International Economy* 5 (1): 36–9.

Schneider, S.H. (1989) *Global Warming*, San Francisco, CA: Sierra Club.

Schwarz, H.E. *et al.* (1990) 'Water quality and flows', in B.L.Turner II *et al.* (eds) op. cit., 253–70.

Seba, D.B. and Prospero, J.M. (1971) 'Pesticides in the lower atmosphere of the northern equatorial Atlantic Ocean', *Atmospheric Environment* 5: 1043–50.

Shaffer, M. (1992) 'Playing God', *The Atlantic Monthly* 269 (4): 8.

Sharpton, V.L. and Ward, P.D.(eds) (1991) *Global Catastrophes in Earth History*,

Boulder, CO: Geological Society of America.

Shaw, R.W. (1987) 'Air pollution by particles', *Scientific American* 257 (2): 96–103.

Sheldon, R.P. (1987) 'Industrial materials – with emphasis on phosphate rock', in D.J. McLaren and B.J. Skinner (eds) op. cit., 374–381.

Shurcliff, W.A. (1986) 'Superinsulated houses', *Annual Review of Energy* 11: 1–24.

Shuval, H.I. (1987) 'The development of water reuse in Israel', *Ambio* 16: 186–90.

Sibley, T.H. and Strickland, R.M. (1985) 'Fisheries: Some relationships to climate change and marine environmental factors', in M.R. White (ed.) *Characterization of Information Requirements for Studies of CO_2 Effects*, Washington, DC: US Department of Energy, 95–143.

Silver, C.S. (1990) *One Earth, One Future: Our Changing Global Environment*, Washington, DC: NAS.

Simberloff, D. (1988) 'The contribution of population and community biology to conservation science', *Annual Review of Ecology and Systematics* 19: 473–511.

Simon, D. (1989) 'Sustainable development: theoretical construct or attainable goal?' *Environmental Conservation* 16: 41–8.

Simon, J. (1990) 'Population panic', *Fortune* 121(11): 160–2.

Simon, J. and Kahn, H. (eds) (1984) *The Resourceful Earth*, Oxford: Basil Blackwell.

Slobodkin, L.B. (1981) 'Listening to a symposium – a summary and prospectus', in M.H. Nitecki (ed.) *Biotic Crisis in Ecological and Evolutionary Time*, New York, NY: Academic Press, 269–88.

Skea, J. (1992) 'Taxes will be insufficient', *Oxford Energy Forum* 8: 8–9.

Smil, V. (1983) *Biomass Energies*, New York, NY: Plenum Press.

Smil, V. *et al.* (1983) *Energy Analysis in Agriculture*, Boulder, CO: Westview Press.

Smil, V. (1984) *The Bad Earth: Environmental Degradation in China*, Armonk, NY: M.E. Sharpe.

Smil, V. (1985a) *Carbon Nitrogen Sulfur: Human Interference in Grand Biospheric Cycles*, New York, NY: Plenum Press.

Smil, V. (1985b) 'Acid rain: critical review of monitoring baselines and analyses', *Power Engineering* 87 (4): 59–63.

Smil, V. (1987) *Energy Food Environment: Realities Myths Options*, Oxford: Oxford University Press.

Smil, V. (1988) *Energy in China's Modernization*, Armonk, NY: M.E. Sharpe.

Smil, V. (1990) 'Nitrogen and phosphorus', in B.L. Turner II *et al.* (eds) op. cit., 423–36.

Smil, V. (1991a) *General Energetics: Energy in the Biosphere and Civilization*, New York, NY: John Wiley.

Smil, V. (1991b) 'Nitrogen and population growth', *Population and Development Review* 17: 569–601.

Smil, V. (1991c) 'Does energy efficiency explain Japan's economic success?' *Current History* 90: 175–8.

Smil, V. (1992) 'China's emissions of greenhouse gases: status and prospects', in T.W. Robinson (ed.) *The Foreign Relations of China's Environmental Policy*, Washington DC: The American Enterprise Institute: 65–72.

Smil, V. (1993) *China's Environmental Crisis*, Armonk, NY: M.E. Sharpe.

Smith, D.J. (1991) 'Fluidized-bed technology probes unit size limits', *Power Engineering* 95 (12): 18–23.

Smith, R.C. *et al.* (1992) 'Ozone depletion: ultraviolet radiation and phytoplankton biology in Antarctic waters', *Science*, 255: 952–7.

Smith, S.E. (1986) 'Drought and water management: the Egyptian response', *Journal of Soil and Water Conservation* 41: 297–300.

Socolow, R.H. (1977) 'The coming age of conservation', *Annual Review of Energy* 2: 239–89.

Solomon, A.M. and West, D.C. (1985) 'Potential responses of forests to CO_2-induced climate change', in M.R.White (ed.) op. cit., 145–69.

Solow, R.M. (1974) 'Intergenerational equity and exhaustible resources', *Review of Economic Studies* Special Issue: 29–45.

Southwick, C.H. (ed.) (1985) *Global Ecology*, Sunderland, MA: Sinauer Associates.

Spencer, R.W. and Christy, J.R. (1990) 'Precise monitoring of global temperature trends from satellites', *Science* 247: 1558–62.

Splinter, W.E. (1976) 'Center-pivot irrigation', *Scientific American* 234 (6): 89–99.

Sprague, G.F. (ed.) (1977) *Corn and Corn Improvement*, Madison, WI: American Society of Agronomy.

SSB (State Statistical Bureau of the PRC) (1991) *China's Statistical Yearbook 1990*, Beijing: SSB.

Stanhill, G. (1985) 'The water resource for agriculture', *Philosophical Transactions of the Royal Society* B 310: 161–73.

Stanhill, G. (1986) 'Water use efficiency', *Advances in Agronomy* 39: 53–85.

Stanhill, G. (1990a) 'The comparative productivity of organic agriculture', *Agriculture, Ecosystems and Environment* 30: 1–26.

Stanhill, G. (1990b) *Irrigation in Israel*, Bet Dagan: The Volcani Center.

Stevenson, F.J. (1986) *Cycles of Soil*, New York, NY: John Wiley.

Steverson, E.M. (1991) 'Provoking a firestorm: waste incineration', *Environmental Science and Technology* 25: 1808–13.

Stinner, D.H. *et al.* (1989) 'In search of traditional farm wisdom for a more sustainable agriculture: a study of Amish farming and society', *Agriculture, Ecosystems and Environment* 27: 77–90.

Stolarski, R.S. (1988) 'The Antarctic ozone hole', *Scientific American* 258(1): 30–6.

Stone, P.H. (1992) 'Forecast cloudy: the limits of global warming models', *Technology Review* 95 (2): 32–40.

Study of Critical Environmental Problems. (1970) *Man's Impact on the Global Environment*, Cambridge, MA: The MIT Press.

Suliman, M.M. (1988) 'Dynamics of range plants and desertification monitoring in the Sudan', *Desert Control Bulletin* 16: 27–31.

Summers, L.H. (1992) 'Summers on sustainable growth', *The Economist*, 323 (7761): 65.

Summers, R. and Heston, A. (1988) 'A new set of international comparisons of real product and price levels estimates for 130 countries, 1950–85', *Review of Income and Wealth* 34: 1–23.

Swedish Ministry of Agriculture (1982) *Acidification Today and Tomorrow*, Stockholm: Swedish Ministry of Agriculture.

Tainter, J.A. (1988) *The Collapse of Complex Societies*, Cambridge: Cambridge University Press.

Takahashi, T. (1987) 'Seasonal variation of CO_2 in the North Pacific Ocean', *Carbon Dioxide Information Analysis Center Communications* Summer: 3–5.

Tansley, A.G. (1935) 'The use and abuse of vegetational concepts and terms', *Ecology* 16: 284–307.

Taylor, A. (1992) 'Can GM remodel itself? *Fortune* Jan 13 : 26–34.

Teal, J.M. and Howarth, R.W. (1984) 'Oil spill studies: a review of ecological effects', *Environmental Management* 8: 27–44.

Thomas, W.L., (ed.) (1956) *Man's Role in Changing the Face of the Earth*, Chicago, IL: University of Chicago Press.

Thompson, T.M. *et al.* (1990) 'Nitrous oxide and halocarbon group', in W. Komhyr (ed.) *Climate Monitoring and Diagnostics Laboratory: Summary Report 1989*, Boulder, CO: US Department of Commerce, 64–72.

Thoreau, H.D. (1854) *Walden*, Boston, MA: Ticknor and Fields.

Toman, M.A. and Burtraw, D. (1991) 'Resolving equity issues in greenhouse gas negotiations', *Resources* Spring 1991: 10–13.

Topfer, K. (1991) 'Request statement: energy and the biosphere', *Environmental Conservation* 18: 291.

Travis, C.C. and Hester, S.T. (1991) 'Global chemical pollution', *Environmental Science and Technology* 25: 814–9.

Tuan, F.C. (1987) *China's Livestock Sector*, Washington, DC: United States Department of Agriculture.

Tucker, C.J. *et al.* (1990) 'Expansion and contraction of the Sahara Desert from 1980 to 1990', *Science* 253: 299–301.

Tudge, C. (1988) *Food Crops for the Future: The Development of Plant Resources*, Oxford: Basil Blackwell.

Turner, B.L. II *et al.* (eds) (1990) *The Earth as Transformed by Human Action*, Cambridge: Cambridge University Press.

Turner, R.E. and N.N. Rabalais (1991) 'Changes in Mississippi River water quality this century', *BioScience* 41 (3): 140–7.

Tutu, D. (1991) 'Do more than win', *Fortune* 124 (15): 59.

Tyndall, J. (1861) 'On the absorption and radiation of heat by gases and vapours, and on the physical connection of radiation, absorption and conduction', *Philosophical Magazine* (Series 4) 22: 169–94, 273–85.

UN (United Nations) (1950–91) *Statistical Yearbook*, New York, NY: UN.

UN (1976) *World Energy Supplies 1950–1974*, New York, NY: UN.

UN (1980–91) *Yearbook of World Energy Statistics*, New York, NY: UN.

UN (1991) *World Population Prospects 1990*, New York, NY: UN.

UN Conference on the Human Environment. (1972)*Conference on the Environment*, Ottawa: Information Canada.

de Unamuno, M. (1921) *Tragic Sense of Life*, London: Macmillan.

UNDP (United Nations Development Programme) (1991) *Human Development Report 1990*, New York, NY: Oxford University Press.

UNDP (1992) *Human Development Report 1991*, New York, NY: Oxford University Press.

UNEP (United Nations Environment Programme) (1982) *Health of the Oceans*, Nairobi: UNEP.

US Bureau of the Census (1975) *Historical Statistics of the United States*, Washington, DC: US Department of Commerce.

USDA (US Department of Agriculture) (1980) *Report and Recommendations on Organic Farming*, Washington, DC: USDA.

USDA (1991) *Agricultural Statistics*, Washington, DC: USDA.

US Department of Commerce (1991) *Statistical Abstract of the United States 1991*, Washington, DC: US Department of Commerce.

US Department of Energy (1987) *Health and Environmental Consequences of the Chernobyl Nuclear Power Plant Accident*, Washington, DC: US Department of Energy.

van Wyk, R.J. (1985) 'The notion of technological limits', *Futures* 17: 214–24.

Vernadsky, V.I. (1929) *La Biosphere*, Paris: Felix Alcan.

Victoria, R.L. *et al.* (1991) 'Mechanisms of water recycling in the Amazon Basin: isotopic insights', *Ambio* 20: 384–7.

Vitousek, P.M. *et al.* (1986) 'Human appropriation of the products of photosynthesis', *BioScience* 36: 368–73.

Wallace, A.R. (1891) *Natural Selection and Tropical Nature*, London: Macmillan.

Wallich, P. and Corcoran, E. (1992) 'The discreet disappearance of the bourgeoisie', *Scientific American* 266 (2): 111.

Washburn, S.T. *et al.* (1989) 'Human health risks of municipal solid waste incineration', *Environmental Impact Assessment Review* 9: 181–98.

Watson, A. *et al.* (eds) (1988) *Air Pollution, the Automobile, and Public Health*, Washington, DC: National Academy Press.

Watson, R.T. *et al.* (1990) 'Greenhouse gases and aerosols', in J.T. Houghton *et al.* (eds), op. cit., 1–40.

Watt, K.F. (1977) Why won't anyone believe us? *Simulation*, 28 (1): 1–3.

Weast, R.C. (ed.) (1990) *CRC Handbook of Chemistry and Physics*, Boca Raton, FL: CRC Press.

Webster, P.J. and Stephens, G.L. (1984) 'Cloud-radiation interaction and the climate problem', in *The Global Climate*, J.T. Houghton, (ed.), Cambridge: Cambridge University Press, 63–77.

Weeks, D.C. (1992) 'The AIDS pandemic in Africa', *Current History* 91: 208–13.

Weeks, J.B. *et al.* (1988) 'Summary of the High Plains regional aquifer-system analysis in parts of Colorado, Kansas, Nebraska, New Mexico, Oklahoma, South Dakota, Texas and Wyoming', *U.S. Geological Survey Professional Paper* 1400–A, United States Geological Survey, Washington, DC.

Weinberg, A.M. (1978) 'Reflections on the energy wars', *American Scientist* 66: 153–8.

Weisser, J.D. *et al.* (1987) 'Sources and production of widely used metals: an assessment of world reserves and resources', in D.J. McLaren and B.J. Skinner (eds) op. cit., 245–304.

Westheimer, F.H. (1987) 'Why nature chose phosphates', *Science* 235: 1173–7.

White, L. (1967) 'The historic roots of our ecological crisis', *Science* 155: 1203–7.

White, M.R. and Hertz-Picciotto, I. (1985) 'Human health: analysis of climate related to health', in M.R. White (ed.) op. cit., 171–206.

Wigley, T.M.L. (1989) 'Possible climatic change due to SO_2-derived cloud condensation nuclei', *Nature* 339: 365–7.

Wilken, G.C. (1991) *Sustainable Agriculture is the Solution, but What is the Problem?* Washington, DC: US Agency for International Development.

Williams, M. (1989) *Americans and their Forests*, New York, NY: Cambridge University Press.

Williams, M. (1990a) 'Forests', in B.L. Turner II *et al.* (eds) op. cit., 179–201.

Williams, M., (ed.) (1990b) *Wetlands: A Threatened Landscape*, Oxford: Basil Blackwell.

Wilshire, H. and Prose, D. (1987) 'Wind energy development in California, USA', *Environmental Management* 11: 13–20.

Wilson, E.O. (ed.) (1986) *Biodiversity*, Washington, DC: National Academy Press.

Winogradsky, S. (1949) *Microbiologie du sol*, Paris: Masson.

Winterbottom, R. and Hazlewood, P.T. (1987) 'Agroforestry and sustainable development: making the connection', *Ambio* 16: 100–10.

Wirth, D.A. and Lashof, D.A. (1990) 'Beyond Vienna and Montreal – multilateral agreements on greenhouse gases', *Ambio* 19: 305–10.

Woodward, J. (1991) 'Enzymes for extracting energy from trash', *ORNL Review* 1991 (1): 38–45.

World Bank (1991) *World Development Report*. New York, NY: Oxford University Press.

World Commission on Environment and Development (1987) *Our Common Future*, New York, NY: Oxford University Press.

REFERENCES

World Energy Conference (1986) *Survey of World Energy Resources*, London: World Energy Conference.

World Resources Institute (1990) *World Resources 1990–91*, New York, NY: Oxford University Press.

World Resources Institute (1992) *World Resources 1992–93*, New York, NY: Oxford University Press.

Yoffe, N. and Cowgill, G.L. (1988) *The Collapse of Ancient States and Civilizations*, Tucson, AZ: The University of Arizona Press.

Young, J. (1990) *Sustaining the Earth*, Cambridge, MA: Harvard University Press.

Zelinsky, W. (1975) 'The demigod's dilemma', *Annals of the Association of American Geographers* 65: 123–43.

INDEX